Property Development

Fifth edition

Sara Wilkinson and Richard Reed

Foreword By Professor David Cadman

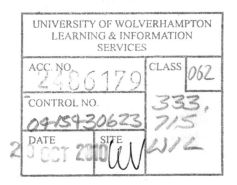

Routledge
Taylor & Francis Group

LONDON AND NEW YORK

First published 1978 by E & FN Spon, and imprint of Chapman & Hall
Second edition 1983, reprinted 1989 (twice)
Third edition 1991, reprinted 1991, 1993, 1994
Fourth edition 1995, reprinted 1996, 1997, 1999
Reprinted by Spon Press 1998, 2001, 2003, 2004, 2006, 2007

Fifth edition published 2008
by Routledge
2 Park Square, Milton Park, Abingdon, Oxon OX14 4RN

Reprinted 2009, 2010

Simultaneously published in the USA and Canada
by Routledge
270 Madison Ave, New York, NY 10016

Routledge is an imprint of the Taylor & Francis Group, an informa business

© 1978, 1983 D. Cadman and L. Austin-Crowe; 1991 D. Cadman,
L. Austin-Crowe, R. Topping and M. Avis; 1995 D. Cadman and
R. Topping; 2008 Sara Wilkinson and Richard Reed, foreword
David Cadman

Typeset in Sabon by
HWA Text and Data Management, London
Printed and bound in Great Britain by
CPI Antony Rowe, Chippenham, Wiltshire

British Library Cataloguing in Publication Data
A catalogue record for this book is available from the British Library

Library of Congress Cataloging-in-Publication Data
A catalog record for this book has been requested

ISBN10: 0-415-43062-3 (hbk)
ISBN10: 0-415-43063-1 (pbk)
ISBN10: 0-203-93342-7 (ebk)

ISBN13: 978–0–415–43062–3 (hbk)
ISBN13: 978–0–415–43063–0 (pbk
ISBN13: 978–0–203–93342–8 (ebk)

Contents

Illustrations

Examples

Figures

Plates

Tables

Foreword

The first edition of *Property Development* was published some thirty years ago not long after one of those market corrections that, despite their inevitability, seem always to catch a new generation of investors and developers by surprise. Indeed, I note that as each new edition has been published, there has been plenty of evidence of the frequency of these ups and downs. Now, with the publication of this fifth edition, we are here again just as another 'unexpected' correction takes place – at least in the UK.

After thirty years and the inevitable transition from the vitality of youth to the tiring of old age, it seems fitting that the authorship of *Property Development* should now move into entirely new hands. And it is a mark of our times that the new authors of this edition should represent the global reach of commercial property markets, markets that I am told no longer have national boundaries.

This new global market is reflected in the ways in which the new authors have sought to extend the text and give new examples of practice. Indeed, it may well be that, eventually, a more assertive approach to globalisation may be needed. In time, a book that is principally about the UK may no longer meet the needs of the next generation of students as they try to come to grips with the markets that will be their working ground. For the time being, however, a focus on the UK with some useful reference to wider markets meets the needs of this book, aimed as it is at students and others that are coming to property development for the first time, and who do so principally from a UK base.

Perhaps the greatest challenge that these students will have to face – and one that we did not even have words for in 1978 – is that of sustainability and, most especially, of climate change. Indeed, it is possible that these matters will create a structural shift in markets of a kind that we have not experienced for some time. This is certainly the place to which my working life has eventually taken me and the more that I study it the greater the size of the challenge appears. Indeed, it is not all certain that we have the skills, the will or the wit to find our way through without much damage.

No longer is property development the protected zone of the landed professions. All sorts of people, now referred to as 'stakeholders', expect to have their say and the policies and actions of developers and investors are expected to be 'transparent'. Property companies and institutional investors have already begun to respond. Indeed, I would say that some at least are leading the field of corporately responsible investment, with increasing evidence of well thought out sustainability policies and standards. They are also beginning to understand how to price the risks of sustainability and climate change. Perhaps this is not so surprising for, as the public sector has abandoned parts of the field that, thirty years ago, were thought to be entirely their preserve and as the private sector has taken occupation, so, too, has the level and scope of their responsibility grown so much so that responsible investment and development is now becoming the established mode.

In all previous editions of this book, I have sought to forecast what might happen next. However, all my experience of trying to foretell the future tells me that uncertainty is ever-present in property markets, the unexpected is to be expected and that, above all else, complacency is the enemy of good investment and development. I think that it is time for me to leave prediction to others. However, being wrong or right in these matters is less important than insisting upon the questions that are raised. I have often said that all forecasts are wrong, the question is: are they useful? Perhaps this should be my epitaph.

David Cadman
London, September 2007

Acknowledgements

The authors would like to thank the following individuals for their input and advice throughout this book, including comments on content of the chapters and contributions to case studies.

- Stephen Brown, RICS Research, London UK
- Paula Judson, Heritage Victoria, Melbourne, Australia
- Miles Keeping, King Sturge, London, UK
- Prof Jon Robinson, University of Melbourne, Victoria, Australia
- Robert Walker, East Lindsey District Council, Louth, Lincolnshire, UK
- John Wombwell, Wombwell Homes, Lostwithiel, Cornwall, UK

1

Introduction

1.1 Introduction

This book introduces and details the processes involved when undertaking property development. It has been written to give the reader a thorough understanding of the nature of property development, what it involves and how to undertake a successful project. At times it is focused on UK practice, although it is truly an international text as frequently noted and commented. This is the fifth edition of the book – since the fourth edition of property development was published in 1995 there have been major changes that have affected the discipline. First, like many other areas of business activity, there has been a substantial move towards globalisation and the property development market is now undertaken on a global as well as national and local scale. For example, Real Estate Investment Trusts (REITs) have been recently introduced into the UK (KPMG, 2007), although they have been available in other global markets, and are an increasingly important source of finance for property developers. Second, the trend towards adopting sustainability within development is affecting the type of property developments perceived as desirable in the marketplace. Finally, property development processes have been significantly affected by advances in technology such as the internet (Dixon *et al.*, 2005), which has speeded up the globalisation of business and allowed best practice in sustainability and property development to be communicated rapidly around the world.

The emphasis in the text is on the practical application of property development, with the reader being taken through all of the stages involved in the process, thereby providing a complete overview. In each chapter a series of discussion points are provided to prompt the reader to reflect on the content of the previous section. Furthermore, a number of illustrative case studies are included to demonstrate the application of the various development stages covered in the chapters. This text is not intended for those who already practice in this field and have experience, but as an introductory guide to students and those new to the field. It is anticipated that upon completion the reader will move onto other texts to deepen and broaden the knowledge base provided here.

The term 'property development' evokes many feelings depending on the particular perspective. The definition adopted in this text is that property development is 'a process that involves changing or intensifying the use of land to produce buildings for occupation'. It is not the buying and selling of land for a profit; land is only one of the raw materials used. Others include the building materials, infrastructure, labour, finance and professional services. For the purposes of this book we are concerned with development that involves building activity for commercial use, whether carried out by the private or public sector.

Property development is an exciting and occasionally frustrating, increasingly complex activity involving the use of scarce resources. It is a high-risk activity that often involves large sums of money tied up in the production process, providing a product that is relatively indivisible and illiquid. Furthermore, the performance of an economy at national and at local levels both directly influence the process. As the development process is frequently lengthy the assumptions made at the outset may have changed dramatically by completion. Success very often depends upon attention to the detail of the process and the quality of the judgement that guides it. Success, however, cannot be judged purely by the size of the profit or loss in financial terms. As a result there are winners and losers whether measured in financial, aesthetic, emotional, social or other terms. For some it can represent an unwelcome change, replacing the familiar with heavily built-up areas fulfilling many functions. For many, however, property development is a worthwhile and very rewarding discipline.

1.2 The process

There are a variety of views on, and descriptions of, the development process. At its most simple, property development can be likened to any other industrial production process that involves the combination of various inputs in order to achieve an output or product. In the case of property development, the product is a change of land use and/or a new or altered building in a process that combines land, labour, materials and finance. However, property development is complex, often taking place over a considerable time frame. The end product is unique, either in terms of its physical characteristics and/or its location. Furthermore, no other process operates under such constant public attention.

The development process may be divided into the following main stages:

1. Initiation
2. Evaluation
3. Acquisition
4. Design and costing
5. Permissions
6. Commitment
7. Implementation
8. Let/manage/dispose

However, property development is not an entirely sequential activity and the stages in the process often overlap or repeat. The sequence is typical of a speculative development where an occupier is not sought until the building has been completed. For example, if the development is pre-sold to an occupier, then stage 8 precedes stages 2–7.

Initiation

Development is initiated when either a parcel of land or site is considered suitable for a different or more intensive use, or if demand for a particular use leads to a search for a suitable site. For the purposes of this book we will focus attention on the main uses such as shops (or retail), offices and industrial. Note many of the same principles also apply to residential developments including low or medium high-density housing in the form of a high-rise building. Often office and industrial uses are combined: buildings suitable for such uses are often defined as business space. The initiative may come from any of the actors or stakeholders in the development process seeking an appropriate site in anticipation of the demand or need for any of the above uses. Alternatively, the initiative may stem from stakeholders anticipating a potentially higher value use for an existing site due to changing demographic, economic, social, physical or other circumstances. In this case, in order to identify the most appropriate use, the initiator will seek to research the market and the potential to obtain the necessary statutory planning consent for the change of use. The roles of the various different stakeholders in initiating the process are examined under 'Main actors' later in this chapter. The initiator may not necessarily be involved in the rest of the development process, depending on their motive or objective.

Evaluation

One of the most important stages of the development process is evaluation as it influences the decision making of the developer throughout. Evaluation includes market research, both in general and specific terms, and the financial appraisal of the proposal (Reed, 2007). The methods of assessing the financial viability of a project are well established and are covered in Chapter 3. Historically, much less attention was given to detailed market research; however, with the application of the internet this has improved and is dealt with in Chapter 7. The process of financial evaluation needs to ensure that the cost of the development is reasonable and viable. For private sector developments, the evaluation establishes the potential for profit in relation to the risks incurred. For the public sector and non-profit-making organisations, it will attempt to ensure that the costs are recovered. An additional objective of the financial appraisal is to establish the value of the site.

This stage of the process should be undertaken prior to any commitment and while the developer retains flexibility. Though the evaluation involves the combined advice of the developer's professional team, the decision to proceed and bear the risk rests ultimately with the developer. It is a continuous process with constant monitoring, relating directly to all the other stages.

Acquisition

Once the decision to proceed is taken, there are many other decisions to be made and steps to be taken before the site can be acquired and the development started. These should include the following listed below.

Legal investigation

Unless the developer is the existing site owner all legal issues concerning the site must be assessed: this includes ownership, existing planning permissions, and any rights of way, light or support. Careful preparation is required to establish who the existing owners of all the rights to the site actually are and what will be necessary to acquire them. Any error in establishing the extent of ownership and the cost or the time in acquiring the rights to the site can seriously affect the viability of the development. The public sector may become involved in the acquisition stage, to assemble a large site with many occupiers and landowners, since they can use their legal powers of compulsory purchase. However, the use of such powers can be both time-consuming and costly. The vast majority of development is undertaken through the co-operation of the original site owners, either by disposing of their interests through negotiation or by becoming partners in the development.

Ground investigation

A thorough physical assessment of the capabilities of the site to accommodate the proposed use should be undertaken. Ground investigations involve the assessment of the site's load-bearing capacity, access and drainage. All existing services (e.g. electricity, water, gas and telephone) should be surveyed to ascertain their capacity to serve the proposed development. If the services are inadequate then the developer needs to assess the cost of their provision or expansion. The investigation should highlight the existence of underground problems such as geological faults and made-up ground, together with the presence of any archaeological remains, contamination, underground services and storage tanks. A site survey must be undertaken to establish the measurement and configuration of the site.

Finance

The developer, unless using internal resources, must also obtain appropriate finance for the development project on the most favourable terms over the entire length of the scheme, before committing to a scheme. The subject of finance will be dealt with in Chapter 4. The developer will normally be concerned with arranging two sorts of finance. Firstly, short-term finance is needed to cover costs during the development process. Secondly, long-term finance (sometimes called 'funding') will be sought to cover the cost of holding the completed development as an investment or, alternatively, to secure a buyer for the completed scheme. The level of detailed information that is required by the providers of the finance varies, but all will require convincing evidence of the ability of the developer, and the soundness of the preparation and appraisal of the scheme.

Design and costing

Design is an almost continuous process running in parallel with the various other stages, getting progressively more detailed as the development proposal increases in certainty. The developer may have detailed knowledge of what design is required if the likely occupier is known or has been secured. In the case of a speculative scheme, the developer may need to work on a number of initial ideas with the agents and the professional team before establishing a design brief for the project. The brief is particularly important for complex schemes as it sets the design parameters for the architect.

Initially, design work will be kept to the minimum to keep costs down prior to developer commitment to the scheme. However, there should be enough detail to enable the quantity surveyor to prepare an initial cost estimate; this in turn is what the developer needs to prepare the financial evaluation. In most cases this means scaled layout plans showing the position of the proposed building(s) on the site, together with simple floor plans showing the internal arrangement of the building on each floor. Plans of the main elevations of the proposed building(s), together with an outline specification of the building materials and finishes, are often desirable. These plans along with the initial cost estimate should enable the developer to prepare the initial evaluation. By the time a decision has been made to submit a detailed planning application for the proposed scheme, the initial plans will be in much greater detail. There will be a full set of plans showing the layout, elevations and section of the building, together with a detailed specification. The developer requires increasing certainty over the cost estimates to improve the quality of the financial appraisal. The quantity surveyor should be able to make a detailed estimate of the building cost at this stage to enable negotiations to commence with building contractors. Care in this preliminary work can save precious time and avoid unnecessary expenses at later stages of the development process.

The design and costing stages include all members of the professional team and continues throughout the construction of the scheme. The developer has to ensure that at each appropriate stage the design and cost estimates are complete to avoid delays to the process. In most cases the final product is very different to the initial design concept, and undergoes many design changes before the final drawings are completed. The developer has to ensure, where possible, significant, and potentially costly, design changes are minimised when the commitment stage is reached.

Permissions

Any development (with a few minor exceptions), which by statutory definition involves a change of use or a building operation, requires planning permission from the local planning authority prior to its commencement. The details of the planning process are dealt with in Chapter 5.

In many instances the developer may, where a building operation is involved, apply for an outline application before full approval is obtained. An outline planning consent establishes the approved use of the site and the permitted size or density of the proposed scheme. The developer only needs to provide sufficient information to describe adequately the type, size and form of the scheme. However, an outline planning consent, on its own, does not allow the developer to proceed with the development scheme; a further detailed planning consent is still required.

A detailed application typically involves the submission to the planning authority of detailed drawings and information on siting, means of access, design, external appearance and landscaping. It is not possible to apply for outline consent for a change of use. There may be a number of outline applications made on a particular site if circumstances change before a developer acquires the site. If the scheme changes, after detailed consent has been obtained, then further approval is required from the local planning authority.

The developer needs to make realistic initial estimates of the likely time and cost of obtaining the appropriate permission during the evaluation stage. The acquisition of planning permission can become complex, requiring detailed knowledge of the appropriate legislation and policies, as well as local knowledge of how a particular planning authority operates. The employment of 'in-house' planners by a developer or the use of planning consultants may be necessary and cost-effective where planning problems are envisaged or encountered. Where permission is refused by the local planning authority the developer may appeal to the Secretary of State.

In addition the developer may be required to enter into a contract with the local planning authority where a 'planning agreement' is negotiated as part of the planning approval. These agreements, which used to be referred to as 'planning gains', deal with matters that cannot be covered as conditions to the planning approval: the provision and maintenance of a public facilities as part of a scheme. For example, there may be improvements to adjacent roads to the site to provide safe access to the site after completion of the development. Planning agreements must be signed before approval is granted and often impose additional development costs, therefore affecting the overall evaluation of the scheme.

In some circumstances there are a variety of other legal consents that may be required prior to commencement of a development. These include listed building consent (the right to alter or demolish a 'protected' or landmark building); the diversion or closure of a right of way; agreements to secure the provision of the necessary services and infrastructure; and, in all cases where building operations are involved, building regulation approval. The prudent developer must clear all legal permission hurdles before making a commitment to the development.

Commitment

A developer must be satisfied that all the necessary preliminary work has been undertaken before any substantial commitment is made in relation to the development. Ideally all the appropriate inputs of land, finance, labour and materials, and the acquisition of statutory permissions must be satisfactorily negotiated before any agreements are signed making the developer liable for any major outlay of money. When the preliminary work has been completed as far as possible, the project must be evaluated once again. This is because it may be that the preparation of the scheme has taken some time and the economic circumstances that determine the success of the development have changed. It is vital, therefore, that the developer pauses for thought until absolutely satisfied that the evaluation is based on the best possible information and the scheme is still viable.

Until the land is acquired, the developer must keep costs to a minimum. The likely costs up to this stage are professional fees and staff time. Depending on individual circumstances some of the professional team may be willing to work on a speculative basis or at reduced fee in order to secure full appointment once the scheme commences. In some cases the developer may be acquiring the land without the benefit of planning permission and, therefore, the contract may be made subject to obtaining the necessary planning approval. In addition, conditional contracts to acquire a site are often entered into when either the developer has had insufficient time to carry out all the important preliminary investigations, or alternatively the developer is yet to secure the necessary finance.

At some point in time all of the contracts to acquire the land, secure the finance and appoint the building contractor together with the professional team will be signed. These contracts may not necessarily be signed together; the developer must aim to achieve this as profits will be maximised. In the case of a non-profit development, ensuring that the commitment is held back until all the resources are in place will minimise cost and risk.

Implementation

The implementation stage can commence once all the raw materials of the development process are in place. At this point there is a commitment to a particular site and to particular buildings at a particular cost spread over a particular time. However, the flexibility, which was previously possible in the earlier phases, has gone. What needs to be emphasised is *the importance of careful evaluation and of maintaining flexibility* as long as possible.

Throughout this stage the underlying goal, and this aspect is often challenging, is to make certain the development is completed within both the time and budget stated in the evaluation, without comprising quality. Depending on the experience of the developer and the complexity of the scheme, this may best be achieved by employing a project manager to co-ordinate the design and building process. The project manager and/or developer must anticipate problems and make prompt informed decisions to minimise delays and extra costs. Furthermore, the developer must take as much interest in the running of the project as in its promotion and the market must be monitored continuously to ensure that the product is right, which may result in amendments to the specification. Where a non-profit development is concerned, the developer must aim to contain costs, while maximising the benefits of occupation. The construction and project management stage of the development process is covered in Chapter 6.

Let/manage/dispose

Although this phase of development occurs often at the latter stages, it must be at the forefront of the developer's thoughts from the initiation of the scheme. In some cases the occupier may have been secured at the start or during the development process. The development's success will depend on the ability to secure a willing occupier at the estimated rent or price, as well as within the period originally forecast in the evaluation. The disposal may take the form of a letting or it may be the outright sale of the freehold interest. In the case of a major retail development there are many lettings, while in that of a single office building the property may be disposed of in one major letting.

It is at the evaluation stage that the letting and/or sales strategy should be thought out, and then subsequently updated, where possible and appropriate, during the course of the development. As such any agent or a member of staff employed by the developer to secure lettings/sales should be included in the development from the beginning of the process. In addition a decision must be made at what point it would be sensible to let or sell the scheme. In many cases it is necessary to complete or virtually complete the development before seeking an occupier. This decision may not be the developer's alone and may be heavily influenced by other actors in the process such as the financiers or the landowner (if they have remained a partner in the development).

At the start of the process the developer has to decide whether the property investment created is to be held as such or sold to realise any profit, unless it has been pre-sold to the long-term financier of the scheme. Such a decision is dependent on the motivation of the developer as well as the prevailing property investment market conditions at the time. However, developers have to be flexible to accommodate any changes in the investment market prior to completion of the scheme. This means that careful thought needs to be given to the investment value at the initial evaluation and design stages. Therefore, if the decision is made to sell the investment to an investor, then the developer needs to fully research their requirements. The location, specification and financial strength of the tenant(s) will be critical in achieving the best price for the investment. The developer may employ an agent to secure a sale of the property to an investor. The agent should be employed as early as possible to advise on the optimal specification and design of the property development scheme.

In accordance with best practice, the development process and the developer's responsibility should not cease with the occupation of the building. There is still a need for the developer to maintain contact with the occupier, even though no direct landlord/tenant relationship may exist. This is because developers can learn more about occupiers' requirements in general and, in particular, the shortcomings of the completed building from a management point of view. Therefore, management needs to be considered as part of the design process at an early stage if the final product is to benefit the occupier and earn the developer a good reputation. The financial success of the development cannot be assessed until the building is complete, let and, where appropriate, sold. Often it may not be until the first rent review under the terms of the letting (typically 5 years after occupation) that the overall picture will become clear.

Discussion point

How do the different stages in the property development process relate to each other?

1.3 Main actors

The development process has been divided into stages and reviewed above. Within each stage, and across some or all of them, there are a variety of important actors who each contributes to the outcome of the property development process and who may have very different perspectives and expectations. The role of the developer has been compared with that of a director of a play who has to manage the diverse and conflicting objectives of all actors on a public stage. As such a director must have the capacity and energy to pull together the performance to ensure it reaches a satisfactory solution.

The actors are considered below in approximately the order they appear in the development process. It is noted, however, that their importance varies from project to project and not all of them appear in every development scheme.

Landowners

Landowners play an important role, whether actively or passively. For example, they may actively initiate development by a desire to sell and/or improve the value of their land. If they are not the active initiators, it may be the case that they become a crucial hurdle to the development, for without their willingness to sell their interest or participate in the development (unless compulsory purchase powers are used), no development can take place. Also the landowner's motivation may well affect their decision to release land for development, and this is the case whether they are individuals, corporations, public authorities or charities. At times they may even take on the role of the developer, either in whole or in part.

Land ownership has been categorised into three broad categories, namely traditional, industrial and financial, as summarised below.

1. Traditional landowners include the church, landed aristocracy and gentry, and the Crown Estate. As such these are significant owners in terms of area and capital value. One of their distinguishing characteristics from other categories is that they are not entirely motivated by the economic imperative. The motives for ownership are wider than return on capital and involve social, political and ideological constraints.
2. Industrial landowners own land incidental to their main purpose, which is some form of production or service provision. This category includes a wide variety of types, including farmers, manufacturers, industrialists, extractive industries, retailers and a variety of service industries. Public authorities of all types, such as centralised and local groups who own land incidental to providing a service or product, might also be included in this category. Consequently the motives of this group are complex, in that they are both dominated and constrained in their attitude to land by their main reason for existence – their product. Furthermore, they may be constrained by their legal status and, therefore, will not always be seeking to maximise their return on land or property in the narrow sense, since that would be seen as subservient to their main purpose. Thus for

them, the economic advantages of releasing land for development are not always evident. If they are forced to sell their land due to a compulsory purchase order, then although they are compensated for the costs of relocating and the disturbance to their business, no allowance is made for the fact that they are unwilling sellers (unless they are residential occupiers). There are intangible losses to commercial businesses that are difficult to value.

3. Financial landowners see their ownership as an investment like any other and may, therefore, be expected to co-operate with development if the return on their land is financially optimal. This category of owners has clear financial motives and is likely to be the most informed type of owner regarding land values and the development process. The major group in this category are financial institutions (pension funds and insurance companies) who own a significant proportion of land by capital value and have been investing varying percentages of their considerable funds in property investment. They may also act as developers directly, or in partnership with developers. Also included are the major property companies who own substantial portfolios of properties and carry out development, and who may, therefore, act as both landowner and developer (see below).

Historically some landowners have had a substantial impact on the spatial layout and the type of development constructed, for example the Grosvenor and Bedford estates in London. Nowadays the planning system has lessened the impact they are able to have on the type of development, but they can still influence the location and planning. The number of owners involved in any particular development also has an important effect, as the greater the number of owners and the smaller their holdings, the more difficult it is to assemble a development site. Many developments have taken years to come to fruition, requiring great patience on the part of the developers.

Developers

Private sector development companies come in a variety of forms and sizes ranging from one-man bands up to multinationals. Their purpose is usually clear: to make a direct financial profit from the process of development – in the same way that any other private sector company operates, whatever their product.

Developers operate primarily as either traders or investors. Most small companies have to trade, that is to sell the properties they develop, as they do not have the capital resources to be able to retain their completed schemes. Many larger public quoted development companies (often referred to as merchant developers) have preferred to trade developments to capitalise on rising rents and values. However, such a strategy can have flaws – for example, in the 1980s many borrowed money on the strength of their future profits and most went into receivership during the financial crash of the early 1990s, the reason being that their limited assets were insufficient to support them. Although some survived, they were effectively controlled by their bankers. Many trader–developers seek to evolve into investor–developers as success enables them to retain profits for investment purposes. At the other end of the scale, some of the largest companies – in terms of capital assets – engage in hardly any new development at all, being content to manage their property portfolio and undertake predominantly refurbishment and redevelopment work. Residential developers operate almost solely as traders as the market is heavily biased towards owner occupation, although many often become significant landowners during the development process.

It also follows that the kind of development undertaken varies considerably. For example, some companies will specialise in a particular type of development, such as offices or retail, and also in particular geographical locations; on the other hand some property developers prefer to spread their risk across types and locations and countries. Some remain in a specialist type of development but cover a wide geographical, and possibly even an international, area. Property companies formulate their policy according to the interest and expertise of their directors and their perception of the prevailing market conditions.

Public sector and government agencies

As a result of central government policy the UK public sector currently undertakes relatively little direct development. Local authorities are primarily involved with developments for their own occupation or community use and the provision of infrastructure. Local authorities are both constrained by their financial resources and limited by their legal powers. Furthermore, local authorities have to be publicly accountable and are obliged to have regard to the overall needs of the community they serve. Local authority involvement in the development process will depend on whether they wish to encourage development or control development in order to maintain standards. Many local authorities undertake economic development activities, with the limited resources they have to promote development and investment in their area. Many active authorities act as a catalyst to the development process by supplying land, and where possible buildings, to achieve economic development of their area. Participation may be limited to the role of landowner in maintaining a long-term interest in a development. Local authorities will often retain the freehold of their development sites and grant a long leasehold interest to the developer, then share in rental growth through the ground rent.

Previously UK government policy was promulgated on the basis of only intervening in the development process where private market forces failed to bring forward development, particularly in areas targeted for economic development. The government's urban regeneration initiatives are administered through several government agencies including Urban Development Corporations (UDCs), English Partnerships, the Welsh Development Agency and Scottish Enterprise. Their role is seen as an enabling role, bringing forward development and attracting investment, in partnership with the private sector. They are able to assist developers with land assembly, site reclamation, the provision of infrastructure, financial grants and, more recently, rental guarantees, following the relaxation of Treasury rules. Other government initiatives aimed at attracting occupiers with financial incentives to specific areas of the country include Enterprise Zones and Regional Selective Assistance. UK government initiatives on urban regeneration are covered in Chapter 2.

Planners

The UK planning system has existed in a comprehensive form since 1947 and is firmly established as the major regulator of property development (see Chapter 5 for a detailed account). Planners can be divided into two broad categories: politicians and professionals. The politicians, usually on the advice of their professional employees, are responsible for approving the development plans drawn up by professionals in accordance with government policy. They are also responsible for determining whether applications for permission for development proposals should be approved or refused. The professionals are responsible for advising the politicians and administering the system.

The main purpose of planning is to 'encourage development' and to prevent 'undesirable development'. The basis for determining planning applications is laid down by statute and a variety of central government policy guidance notes. Local government must adhere to these and determine its own local policy through the main medium of development plans. Individual planning applications are determined in the light of these development plans, written government policy and advice, previous decisions and the particular nature of the application. Nevertheless often in practice there are many gaps and conflicts in the guidance, which means that developers often employ planning consultants to assist them in negotiations with planners. Developers need to know what use, what density and what design standards are required in order to obtain permission. A successful application is usually best achieved by prior negotiation with the authorities and this may involve agreement by the developer to provide infrastructure or community facilities in the case of a large development, known as a 'planning agreement'. This type of agreement is endorsed by government guidance provided it reasonably relates to the development proposed. In the context of tight public spending controls, a planning agreement is seen by local authorities as a means of securing useful benefits for the community. However, the issue of planning agreements has been controversial and there is a limit as to how much a developer can afford and so a test of 'reasonableness' is usually applied to applications.

Planning authorities differ widely in their policies towards development. Those in areas of low economic activity typically wish to encourage development activity, putting only minimal restrictions on proposals, particularly those that will provide employment. Authorities in areas of high economic activity mainly see their role as imposing higher standards, and even slowing down development in order to achieve a better balance of uses and improved design of buildings. In this situation there is an increased level of conflict between developers and planners, leading to increased use of the appeals system, especially in areas of high economic activity. In some instances the conflict is caused by the politicians ignoring the advice of the professionals.

Financial institutions

Unless a development is being financed entirely with a developer's own capital or that of a partner, then financial institutions, as providers of finance, have a very important role in the development process. Financial institution is a term usually used to describe pension funds and insurance companies. However, there are many other financial intermediaries such as clearing and merchant banks (both UK and international), as well as building societies who also provide finance for property development.

There are two main types of money required for development: short-term money, also known as 'development finance', to cover the costs during the development process; and the long-term money, or 'funding', to cover the cost of holding the completed development as an investment. Alternatively, the developer will in the long term seek a buyer for the completed scheme to repay the short-term loan and realise any profit.

Financial institutions (pension funds and insurance companies) are motivated by direct financial gain. However, unlike developers, they take a long-term view, needing to achieve capital growth to meet their payment obligations in real terms to pensioners and policyholders. Note that pension, life and investment funds are usually judged on their short-term performance, both in relation to other forms of investment and to the returns they achieve against competing funds. They seek to minimise risk and maximise future yields. The yield on any investment is the annual income received from the asset expressed as a percentage of its capital cost or value. Property or real estate is only one of a number of investments the institutions invest in and may represent only 5–15 per cent of their entire portfolio of investments. In the case of property the financial institution will receive a lower initial income when compared to a fixed-interest investment, but this will be more than compensated by the long-term growth.

Both short-term and long-term finance may be provided to a developer by what is called 'forward-funding' a development; in other words they agree to purchase the development on completion whilst providing all the finance in the interim. Almost all of the risk passes to the developer who will in most cases provide a financial guarantee. Alternatively, they may act as developer themselves to create an investment: all the risk is theirs but they do not have to provide a profit to the developer. Some only purchase completed and fully let developments since they perceive the overall development as being too risky. In order to be persuaded to take on the risks associated with development, rather than purchasing a completed and let scheme, they need a higher return or yield.

Whether acting as developer, financier or investor they tend to adopt rigid and conservative policies, although they all differ in their individual criteria. However, they all tend to seek a balanced portfolio of property types rather than specialising in one particular use. Most try to spread their investments geographically. They will seek properties or developments that fit their specific criteria in terms of location, quality of building and tenant covenant (i.e. financial strength). As a result developers sometimes lean towards to developing schemes in accordance with the financial institutions' specification rather than that of the occupiers. Financial institutions wish to purchase a building that has the widest tenant appeal, consequently their advisers may take a conservative view and recommend the highest specification, which can lead to less sustainable and over-specified buildings.

The developer may approach the banking sector for funding if a development is either not 'institutionally acceptable' or if the developer is not prepared or is unable to provide the necessary guarantees. Alternatively, the developer may prefer to use debt finance in a period of rising rents and values to maximise the potential profit on completion. There are a great variety of methods of obtaining finance from the banks both for short- and medium-term finance (these are covered in detail in Chapter 4). The banks also aim to make a financial profit from the business of lending money. Bank lending may take the form of 'corporate' lending to the development company or lending against a particular development project. The banks will use the property assets of the company or the property as security for the loan. Property is attractive as security as it is a large identifiable asset with a resale value. Banks wish to ensure that the proposed development is well located, the developer has the ability to complete the project and the scheme is viable. With corporate lending the bank is concerned with the strength of the company, its assets, profits and cash flow. Depending on the size of the loan, and in particular where the bank is exposed to above normal risk, then the banks may secure an equity stake in the scheme.

Residential developers who build housing for owner-occupation only require short-term development finance, which is usually provided by the banks. Their ability to raise finance is based on their 'track record' and the value of the scheme.

For public sector development, the sources are similar but much more difficult to obtain, primarily due to tight central government control on public sector borrowing. Some local authorities may obtain funding through grants for urban regeneration projects in specific geographic areas, mainly from central government sources and European Structural funds. However, access to such funding is subject to competition and targeted at schemes carried out in partnership with the private sector and the community.

Developers may also obtain financial assistance from the various government agencies in the form of grants, rental guarantees and equity participation through the provision of land. However, the developer has to prove that the project would not proceed without such assistance and, importantly, that it will create jobs relevant to the local population.

Building contractors

Building contractors are employed by developers to construct the development scheme and their prime objective is direct financial gain. There are a great many building contractors and a considerable variety of contractual systems for obtaining a completed building (see Chapter 6 for a detailed account).

Some development companies employ their own contractors. Residential developers tend to employ all the necessary expertise in-house: this is why they are often referred to as house-builders. Development companies may keep their contracting division at 'arm's length' as an entirely separate profit-making centre though there is a general trend towards a more integrated approach. A builder who takes on the role of developer, e.g. house-builder, also takes on the additional risk associated with the development process. When a builder is merely employed as a contractor the financial profit is related to the building cost and length of contract. Under a 'design and build' type of contract a contractor will take on a design role that will involve a greater element of risk in relation to the responsibility for cost increases. Larger contractors with the relevant expertise may take on the role of a management contractor and manage all the various subcontracts for the developer in return for a fee. In the case where the builder is the developer, then a larger return is required due to the risk involved, but by combining the building and development profit an overall lower profit is acceptable. For builders who employ a substantial labour force, an additional objective may be to ensure continuing employment. Sometimes this can only be achieved by cutting tender price, leading, therefore, to decreased profits.

Essentially, building contractors carry out a specialist activity within the development process, commencing at a time of maximum commitment and risk for the developer. The prudent developer will, therefore, ensure the capability and capacity of the contractor(s) to undertake the proposed work, seeking the right balance between the lowest tender and quality of performance. However, it is not in the contractor's or developer's interest to have a situation where the contractor is unable to make a reasonable profit from the scheme. Clearly it is not in the developer's interest for the contractor to go bankrupt or to be compromising on quality.

Finally there has been a major change in the way public infrastructure can be delivered in the UK. Public Private Partnerships (PPPs) are a key component of the UK government's strategy. PPPs are developed with three objectives:

1. to deliver significantly improved public services, by contributing to increases in the quality and quantity of investment;
2. to release the full potential of public sector assets to provide value for the tax payer and wider benefits for the economy;
3. to allow stakeholders such as the users of the service, taxpayers and employees to receive a fair share of the benefits of the PPP.

For mutual benefit, PPPs bring together a public body and a private company in a long-term joint venture for the delivery of high-quality public services. The UK government has realised that the public sector cannot always deliver major investment projects, although the private sector may bring about benefits including:

- increased efficiency and the increase in innovative ways of delivering public services;
- motivation to invest in high-quality assets to optimise maintenance and running costs;
- improved management of the risks involved in completing complex investment projects to time and budget, and providing quality services thereafter.

Drawing on the best of the public and private sectors, PPPs provide additional resources for investment in public sectors and the efficient management of that investment. PPPs cover a wide range of different types of contractual and collaborative partnerships, such as the Private Finance Initiative (PFI), the introduction of private sector ownership into state-owned businesses, the sale of government services into wider markets and the generation of commercial activities from public sector assets through, for example, the 'Wider Markets Initiative'.

Agents

Commercial agents, or estate agents where residential developments are concerned, may be instrumental in initiating the development process and/ or bringing together some of the main actors in the process. They also form the link between the developer and the occupier, unless the developer uses in-house staff to perform the agent's role; in this case the occupier is not represented by an agent. Therefore, they play a very important role within the development process and are often involved in every stage of the process. Agents are able to perform this role due to their detailed knowledge of both the property market in terms of demand and current rents/prices, relying upon their 'personal' contacts with developers, occupiers, financial institutions and landowners. This emphasises the fact that the development industry is primarily all about 'people'.

The agent's aim is to make a direct financial profit from the fees charged to their client (be it the developer or occupier) for carrying out a professional service. In the case of introducing one party to another they only receive a fee if the transaction is completed (e.g. the property is let) but it is nearly always related to the value of the transaction in percentage terms. They may be instrumental in initiating the development process, either by finding a suitable site for a developer or advising a landowner to sell a particular site due to its development potential. Unless they are retained by a developer to specifically find sites to suit a particular use, they take the initiative in identifying suitable sites and 'introduce' them to developers. In addition they will introduce sites to those developers who they consider have the appropriate expertise and resources to both acquire the site and complete the development. The agent will negotiate with the landowner on the developer's behalf and then advise the developer on all matters relating to the evaluation stage. In return they may secure not only a fee for introducing the site (usually related to the purchase price) if the acquisition stage proceeds, but they may also secure appointment as post-completion letting and/or funding agent for the development scheme. Even though it can often be a lengthy and time-consuming process for the agent, the rewards can be high. If an agent acts for a landowner then they advise on both the likely achievable land value and the likely market for the site.

Agents may be used by developers to assist them in securing the necessary finance for a development scheme due to their knowledge of the requirements of the financial institutions or banks. Furthermore, many institutions retain agents to advise them generally on their property investments including development funding: they may specifically find development opportunities for their client to fund. The institution's agent will normally advise their client throughout the process and be one of the letting agents on the scheme. Some of the larger or more specialist firms of chartered surveyors, with financial service divisions, may act as financial intermediaries arranging funding packages with banks and other institutions in return for a fee related to the size of the loan.

Agents are widely employed by developers as letting or selling agents, where they provide that all important link between developer and occupier. In performing this role they should be involved from the start of the development to enable them to advise the developer on the occupier's viewpoint. However, unless they have a specialist marketing department they usually cannot give total comprehensive advice on the precise requirements of the market. A developer will need to commission market research to obtain more detailed knowledge of the specific market for the completed development. Some developers may employ an in-house team, although the advantage of the agent is their knowledge of the market in general and their contacts with potential occupiers or their agents.

Developers, landowners, occupiers, financiers and property investors may at some stage in the development process employ a suitably experienced qualified chartered surveyor or valuer to undertake professional work to assist their decision-making process. Chartered surveyors and valuers are employed by the vast majority of commercial and residential agents to enable them to undertake professional work alongside their agency work. Professional work that is related to development may include valuation, building surveys and management. Developers will require an independent valuation to check their own opinion of value, particularly where they have insufficient knowledge themselves or it is required by a financier of the development. Independent and in-house valuers are also used by financial institutions and banks on schemes for which they are considering making loans or granting mortgages, including the asset value of any security being provided by the developer. Financiers will often employ building surveyors to check on the construction phase of the development to ensure that it is being built to the right specification, as well as to certify drawdown of the development loan. In the public sector, local authorities, central government and the Inland Revenue all use valuers to advise and check on any development-related work.

Professional team

The development process is complex and, therefore, the majority of developers do not have the skills or expertise to carry out a major development. As a result, developers employ a range of professionals to advise them at various stages of the development process depending on their needs; these include the following listed below.

Planning consultants

Planning consultants negotiate with local planning authorities to obtain the most valuable permission for a development, particularly with large or sensitive schemes. If a planning application is refused they will be employed to act as expert witnesses in presenting the case for the developer. Furthermore, planning consultants can advise landowners to ensure that the sites within their ownership that are allocated under the development plan are aligned to their most appropriate or valuable use. This may involve negotiations with the local planning authority at plan preparation stage, or subsequent representations at an inquiry into the development plan. In performing this role the planning consultants can be important initiators of the development process.

Property economics consultants/valuation surveyors

Property economic consultants or valuation surveyors are employed at the critical evaluation stage to provide a detailed analysis of the characteristics of the market in terms of the underlying demand and competitive supply. Many financiers, particularly the financial institutions, insist on market analysis when considering a development funding proposal. In addition, they often employ researchers in-house in formulating their funding criteria and policy (see Chapter 7 for further details on the use of research).

Architects

Architects are employed by developers to design the appearance and construction of new buildings or the refurbishment of existing buildings. They may also administer the building contract on behalf of the developer and certify completion of the building work. In the case of refurbishment work, building surveyors are usually employed to survey the existing building and advise on the alterations and provide contract administrations services. Architects are normally responsible for obtaining planning permission where a planning consultant is not employed. With a refurbishment the building surveyor will perform this task. Architects are normally paid on a 'fee' basis, usually a percentage of the total building contract sum.

It is important that the architect is employed at the earliest possible stage to ensure that all design work is ready when construction is due to commence. It is also important to employ architects with the appropriate experience, reputation, resources and track record. A developer should ensure the architect has the right balance of skills to produce not only good architecture, but also a cost-effective and workable design attractive to occupiers. This balance is very often hard to achieve and, therefore, it is important for a developer to produce a clear architectural brief from the start. Problems start when there is insufficient communication between architect and client.

Some firms of architects offer a comprehensive service, including project management, engineering and interior design work. This may be effective on some development schemes but most developers tend to prefer to assemble their own professional teams to achieve the most effective and experienced team. Some development companies, particularly those which specialise, employ in-house architects and design professionals.

Construction economists or quantity surveyors

Construction economists or quantity surveyors are, in simple terms, 'building accountants' who advise the developer on the likely costs of the total building contract and its associated costs. Their role can include costing the designs produced by the architect, administering the building contract tender, advising on the most appropriate form of building contract (procurement), monitoring the construction and approving stage payments to the contractor. They are increasingly becoming involved in the administration and management of design and build contracts. Like architects, their fee is based on a percentage of the final contract sum. The choice of quantity surveyor should be based on appropriate experience and reputation. Importantly, the developer should select a quantity surveyor who works well in partnership with architects and other members of the professional team to produce cost-effective designs. Also, a good quantity surveyor should be able to provide the developer with cost-effective ideas as alternatives to those proposed by the architect.

Engineers

Structural engineers are employed to work with the architect and quantity surveyor to advise on the design of the structural elements of the building. They will also participate in the supervision of the construction of the structure. Civil engineers will be employed where major infrastructure works and/or ground work is required. On large and complex schemes, often mechanical and electrical engineers are used to design all the services within the building. Engineers are usually paid a percentage fee based on the value of their element of the building contract.

Project managers

Project managers are employed to manage the professional team and the building contract on behalf of the developer. Project managers are normally only employed on the larger and more complicated schemes. They often come from other built environment professions – for example they may have been educated and trained as a quantity surveyor or civil engineer or architect initially.

Developers often act as project manager or rely on in-house staff or another member of the professional team. They should be appointed before any of the other professional team or the contractor so that they are in a position to advise the developer on the best professional team for the project. Their fees are normally based on a percentage of the building contract sum, but often there may be an incentive for managing the scheme within budget and on time. Developers often perform the role of project managers themselves for occupiers who wish to employ the expertise of a developer in constructing their own premises.

Solicitors

Solicitors are needed at various stages throughout the development process, starting with the acquisition of the development site through to the completion of leases and contracts of sale. They are often involved with the legal agreements covering the funding arrangements entered into by the developer. If the developer has to appeal on a planning application then both solicitors and barristers may be involved in presenting the developer's case at an inquiry. With some schemes collateral warranties are required by purchasers and the solicitors will be involved in this process. Collateral warranties are defined as agreements under which parties with contractual obligations, in connection with construction or operation of a project, accept liability to the lenders for their performance.

Accountants

In some cases specialist accountants may be employed to provide advice on the complexity of tax and Value Added Tax (VAT) regulations that can have a major cost impact on a development. They may also provide advice on the structure of partnership or financing arrangements.

Note that the above is not intended as an exhaustive or detailed list of the various professionals and specialists employed during the development process or their professional expertise or their roles. However, it is designed to give an insight into the main professional support. There are a considerable variety of other specialists who may be necessary depending on the circumstances of the project and its complexity. For instance, other relevant professionals may include highway engineers, landscape architects, land surveyors, soil specialists, archaeologists, public relations consultants and marketing consultants. Hence the above shows the variety of skills that are required within the development process.

Objectors

There are two categories of objectors who can potentially cause delay and possible abandonment of development projects. The first may be purely 'amateurs' and self-interested neighbours of the proposed development.

They are sometimes referred to as 'NIMBYS' ('not in my back yard') and, where organised, they can achieve considerable obstruction to the progress of development proposals. For example, in the past such well-organised and influential organisations contributed towards the abandonment of proposals for private sector new towns in Oxfordshire and Hampshire.

The second category is the well-organised professional, permanent bodies at local and national level. Locally they include amenity societies who take an interest in every proposal affecting their local environment and heritage. Nationally they include the Victorian and Georgian societies. These bodies have considerable influence with the local planning authorities and tend to always be consulted on major applications. There are also official quangos (quasi non-government organisations), such as the Royal Art Commission, the Nature Conservancy Council and English Heritage, who may take an interest in important buildings and existing flora and fauna. These organisations are well informed and have a good knowledge of the planning and development processes.

Finally, another organised group that emerged during the 1990s were the environmental activists. To date their activities have been focused on larger developments. During the early 1990s environmentalists occupied the Twyford Down section of the M3 extension in Hampshire causing major delays and unwanted media attention to the project. Such protests are not confined to the UK and occur in other countries such as the Three Gorges Project in China or the Tehran–Shomal (Northern Iran) highway. The protests can take the form of direct action such as occupation, or a more passive approach such as letter writing to officials.

Developers must be aware of their interest (or likely interest) and be prepared to accommodate or refute their opposition. Ideally such negotiations should be carried out before a planning application is made to avoid lengthy delays. Opposition can be costly to a developer, either by imposing higher standards and costly alterations or lengthy delays. At worst, opposition can lead to the complete abandonment of proposals that may have been sound. Prudent developers need to account for objectors when evaluating their proposals.

Occupiers

Unless the occupier of a building is the developer or is known early in the process, then the occupier is not regarded as a main actor within the development process as they are often unknown until the development is complete and let/sold. Their demand for accommodation triggers the development process and influences both land prices and rents, to which developers respond (see 'Economic context'). The occupier is an actor within the process and their requirements should be researched at the beginning of the process. Developers in the past have tended to produce buildings in accordance with the requirements of the financial institutions and the needs of the occupier have been overlooked. However, there is a growing recognition within the industry that this has to change and many developers now commission research into occupiers' requirements at a general and specific level (see Chapters 7 and 9).

Of course when the occupier of the scheme is known early on, then the occupier becomes the most significant actor in the process. The building will be constructed in accordance with the occupier's requirements, which can be very specialised, particularly with industrial users. The developer may need to persuade the occupier to compromise on their requirements to provide a more standard and flexible type of building, so that the investment market for the building is wider in the event of disposal. The developer will also be concerned to protect the value of the building as security for loan purposes.

Occupiers, as commercial businesses, largely tend to regard the buildings they occupy as an overhead incidental to their activities as providers of a service or product. Although some companies do employ an in-house property team (consisting of surveyors and facilities managers) and many are set up as a profit-making centre, in some instances acting as consultants to other occupiers (e.g. Digital's facilities management consultancy), many occupiers tend to fail to adequately plan their property requirements far enough in advance. They simply react to changes in their business as they happen. The property requirements of occupiers are influenced by both the short-term business cycle and long-term structural changes underlying the general economy (see 'Economic context' and Chapter 7). Either of these factors may influence occupiers at a specific level or across the business sector in which they operate. Their demand for accommodation is also influenced by advances in technology affecting both working practices and their physical property requirements.

Both agents and developers criticise occupiers for not knowing what they want, although many companies are becoming far more knowledgeable about the role of property within their businesses and their requirements in terms of specification. This is evidenced by the expanding profession of facility management. Occupiers have differing requirements and priorities, particularly in the case of offices, making the task of the developer difficult in producing a building suitable for as many tenants as possible. The response of the financial institutions is to seek the highest quality specification with a layout to suit the widest possible range of tenants. As a consequence an occupier may be forced to occupy a building that compromises their requirements in terms of location or specification. There have been many examples of occupiers stripping out buildings and refitting them in accordance with their requirements. For example, it is impossible to achieve a multipurpose building, say in the City of London, where financial service companies require large open plan floors with high ceilings and professional services companies require cellular offices with good natural daylight.

A further area of conflict between occupiers and the financial institutions concerns lease terms. Depending on the local market conditions of supply and demand, occupiers are demanding more flexible lease terms to enable them to respond in the short term to their property requirements. Although gradually changing, many financial institutions prefer the traditional 25-year lease term with upward-only rent reviews. However, many will accept shorter lease terms (from 15 years) and options to break if the market conditions require such agreements. In countries other than the UK is it common to find quite different lease terms, for instance in the US where 5-year leases with an option to renew for another 5-year term is the 'norm' for commercial office space.

Developers and the financial institutions are taking more account of the needs of occupiers. There is recognition that the office buildings of the 1980s were over-specified and that reducing the costs of occupying a building are important to occupiers. Note that the British Council for Offices (BCO) has published guidelines for best practice for office buildings. The trend is towards energy efficient buildings with the maximum use of natural ventilation and the use of air conditioning only where absolutely essential (see Chapter 9).

Discussion points

Who are the main actors in the property development process?
For a large-scale, out-of-town retail development who do you think would need to be part of the developer's team and why?

1.4 Economic context

Property development does not, of course, occur in a vacuum. Occupier demand is a reflection of the short-term and long-term changes in the economy. Availability of development finance is also linked to conditions in the wider economy. The economic context is important to developers in both a specific way (in so far as the local economic context helps to determine the market for an individual scheme) and in a more general way via the wider economy (as it affects general property market conditions and the confidence of occupiers, investors and developers).

The local economy

However, local economies are vitally important and research studies show that most demand for an individual office or industrial development is drawn from a small geographical area around the scheme. Similarly, the ability of a retail scheme to attract national retailers depends crucially on the spending capacity of the local population. Since local economic conditions will help determine how much development, and of what type, is appropriate in a particular place, it is in the interest of any developer to look beyond the individual scheme to the wider economy. The local economic context can thus be a useful indicator of the likely viability of any development project, and can be used alongside the development appraisal. The role of market research is covered in Chapter 7.

The national economy

At a more general level, the interaction of the short-run business cycle with property cycles creates great variability in a developer's plans, and the ability to progress schemes at different times. The developer also needs to be responsive to the more evolutionary changes that occur in occupier preferences as a result of long-term changes in the structure of the economy.

The business cycle

Economic and property research has established the nature of the link between the economic, or business, cycle and the property market. Useful early references on this complex topic include the report and papers on building cycles by Richard Barras (1994) and reports from the Royal Institution of Chartered Surveyors on property cycles, carried out by the University of Aberdeen and the Investment Property Databank (1994). Three important cycles were identified, all of which exhibit different periodicity – the business cycle (which drives the occupier market), the credit cycle (which influences bank and institutional funding) and the property development cycle itself. A simplified version of Barras' analysis is provided below starting with the business upturn:

1. Strengthening demand, rising rents and capital values trigger the start of the new development cycle upswing.
2. If credit expansion accompanies the business cycle upswing, it can lead to a full-blown economic boom. The banks may also fund a second wave of speculative development activity.
3. However, because of the long lead times in bringing forward new development, supply remains fairly tight and values continue to rise.
4. By the time the development cycle reaches its peak, the business cycle has already moved into a downswing, accompanied by a tightening of monetary policy to combat the inflationary effects of the economic boom.
5. As the economy subsides, the demand for property declines; rents and values fall as a result and the vacancy stock increases in supply.
6. As the economy moves into recession, the fall in rents and values continues, property companies are hit by the credit squeeze, bankruptcies increase and the development cycle is choked off.

Barras (1994) noted the UK experience was such that a property boom was typically followed by a more muted development cycle. At this point lenders and investors struggled with debts incurred during a downturn or recession and were then unable or disinclined to fund speculative development. At the same time, oversupply of property built in the previous boom was sufficient to meet demand during the whole of the following business cycle. It is only when supply is exhausted, at the start of the next business cycle upswing that the speculative development cycle takes off once more. This is the reason that absorption and vacancy rates are so important to market analysts. Absorption rates are the speed at which vacant space is taken up by the market. Clearly, the likely success of development projects will be influenced by where in the cycle they are started and completed. However, recent experience over the last decade has been for a sustained period of growth in the global economy as well as national economies like the UK, so much so that some commentators are suggesting that the old so called 'boom–bust' economic cycles are gone. It is true to say that many, i.e. those who entered the workforce after 1995, have not experienced an economic downturn in their working lives.

Structural change

Underlying the short-term business cycle are longer-term shifts in occupier requirements that result from structural changes in the economy. For example, the expansion of demand for very large warehouses or so called 'big box' retail space has resulted from strategic reorganisation within the retail and logistics industries, helped by information technology. Similarly, the changes in working practices amongst office occupiers, again encouraged by developments in information technology, will most likely generate demand for new kinds of office building in the future. The term 'hot-desking' was given to the sharing of desk space by more than one employee and is a phenomenon of the late 1990s. Similarly the growth of the telecommuter (people working from home using IT has created demand for study space within residential property). Developers (and investors) who monitor these long-term changes can begin to create new types of product ahead of the rest of the market; equally they can avoid being left with buildings that have a diminishing 'shelf life'. This is where property market research is useful, a subject covered in Chapter 7.

The global economy

The uptake of technology around the world has allowed our environment to be classified as a global one as business became able to market products and services all over the planet. Technology has enabled business to develop partnerships and alliances globally. Prior to globalisation the USA dominated the global economy, however, their share has reduced to around 25 per cent; the trend is expected to continue as the economies of newly industrialised countries, like China, expand at a faster rate. With the global economy and use of the internet, people and businesses are aware that they are competing with others around the world for contracts and business deals.

With the globalisation of business, the economic performance in some countries such as China and the US can substantially affect other regions such as Asia or Europe, and this may need to be taken into account. The last decade has seen a substantial increase in investment in properties from overseas buyers. For example, in Europe there has been an increase in the purchase of investment properties in Turkey and the Adriatic by Northern Europeans, whereas in Australia there is a considerable Asian market buying into investment property in the major cities. Property developers could be affected considerably by changes in the global economy and need to be aware of prevailing trends and forecasted changes.

Discussion point

How has the economic context that property developers operate within changed over the past decade?

Reflective summary

This chapter has contained an overview of the very complex activity of property development, reviewing it as a series of stages involving many actors with differing objectives, operating within the overall context of the building cycle and its interaction with the business and credit cycles. Development may be initiated by any of the main actors identified but it can only take place with the consent of the landowner (unless compulsory purchase powers are used). As a development proceeds through the various stages the developer will become increasingly committed and flexibility will be reduced, exposing the developer to greater risk. Before developers make a commitment to both acquiring land and signing a building contract they should obtain all the necessary consents, carry out all the necessary investigations and secure the necessary finance. In addition, a thorough financial and market evaluation should be carried out with the best information possible to establish the project's viability and the occupier market. The success of a development may often depend on an element of luck as well as the developer's judgement and skill, depending on the interaction of the building and business cycle on completion.

Land for development

2.1 Introduction

Land acquisition is often the developer's first major commitment to a project. The initiation, evaluation and acquisition stages of a development are closely linked and frequently occur simultaneously. However, the acquisition of land should not be completed until after the evaluation process has been completed (see Chapter 3). This chapter examines the initiation and acquisition stages of the development process.

Site selection is vital to success as it affects the nature of a project. If the site has a poor location or there is a lack of demand for the accommodation irrespective of location, no amount of good design or promotion can overcome this disadvantage to a project. Land is unique and each site has its own characteristics. The site acquisition process can be frustrating and unpredictable since many factors, often outside the developer's control, can influence its success.

This chapter looks at the methods that developers use to identify and acquire sites. The role of landowners as initiators of the process is described, as well as the role of local authorities and government agencies in bringing forward land for development and attracting private sector investment. It includes an examination of the taxes and grants available to the developer in the areas of the country where development schemes might not occur if left to market forces. Whilst this chapter uses examples from the UK to illustrate points, readers outside of the UK can apply the principles of supply and demand to land use to the country in which they are located.

2.2 Site identification

The first step in finding a development site is to establish a strategy defining the aims, nature and area of search. The starting point may be the development company's own business plan, which sets out the aims of the company. As discussed in Chapter 1, development companies may restrict themselves geographically or to particular types of development, e.g. offices or residential. The overall strategy and aims of a development company drives the identification of sites and development opportunities, and within this overall strategy a developer has to define closely specific areas of search and their exact requirements, especially in relation to the size, nature and location of sites. The geographical area of search for sites is related to factors such as:

- the location of the developer's own office or offices;
- rhe desire to spread risk across a number of locations;
- the availability of development finance;
- the results of market research.

The location of the company's offices is important because the further the site is away from the office, the more likely that the management of the project will be less effective. If the developer operates within a local area they tend to establish good contacts with agents, occupiers and the local planning authorities. When a development site is a considerable distance from the developer's office, then a good working relationship will need to be established with local agents and it may be wise to undertake the project with a local partner. Local knowledge is invaluable. However, some development companies, particularly the larger ones, may prefer to distribute their exposure to risk by spreading their development activities over different locations. Therefore, if an oversupply of accommodation occurs in one location it will only impact on some projects and not the entire development portfolio.

A developer's plans for the identification of sites may be influenced by the way in which the developer finances development projects. For example, if a developer seeks finance from a financial institution then the developer needs to be aware of the institution's preferred locations for property investment. The requirements of financial institutions are discussed in detail in Chapter 4. During the late 1980s development boom, the location of speculative development sites became less critical for the purposes of obtaining finance, with the widespread availability of bank finance and the decline in the dominance of the financial institutions as funders in the development market. In any market conditions, development funding from either banks or financial institutions is likely to be more successful on well-located 'prime' sites, as projects built on such sites have the widest tenant appeal and reduce developer risk. Whatever the state of the property market, the prudent developer should always seek the best locations appropriate to the proposed use. A secondary location represents a higher risk to the developer in terms of achieving a letting/sale and securing funding, which should be recognised in the evaluation of the scheme (see Chapter 3).

Still, the developer's perception of occupier demand, backed up by market research, should be the most important factor in influencing the area of geographical search. The developer's skill and knowledge is important in identifying areas of potential growth where market forces will provide demand for accommodation that will exceed supply by the time a development project is on stream. Forward-thinking developers may commission research at a strategic level to identify trends in the market and areas of potential opportunity (see Chapter 7 for strategic research). Whilst recognising the risks involved, developers seek opportunities to be ahead of the market. Road and public transport improvements with major redevelopment schemes are some of the more obvious factors that may influence the demand for accommodation in a particular area. Developers will seek to purchase a site in a town that is likely to benefit from a planned major road scheme. For example, many developers bought sites in towns on London's M25 orbital motorway as its route became known in the early 1980s. Market research should seek to identify the current and projected levels of supply and demand of various types of accommodation in a particular area. Developers should examine trends in rental and capital values, and the underlying economy in the areas of search identified. The use of market research in formulating a land acquisition strategy is discussed in greater detail in Chapter 7.

As part of any market research the developer should identify the factors that influence occupiers in their choice of location. Various factors influencing the choice of location for each type of commercial development are discussed below.

Office/'business space' development

The majority of office development tends to take place in London, the East of England and the South East of England. Outside this area, office development activity is concentrated in the major cities and regions, e.g. Bristol, Birmingham, Manchester, Leeds, Edinburgh and Glasgow. According to government statistics, the total amount of office space in England in 2005 totalled 97.875 million m^2, of which 29.1 per cent was in London.

Road, rail and air communications are vital in the consideration of locations for office development. In London, and Britain's other main cities, the proximity to good public transport is important for office locations in the central areas. In relation to office development in provincial towns and 'out-of-town' business or office parks, proximity to the national motorway network and airports are important. This was shown in the South East with the growth of towns along the motorways (M3, M4 and M25) and near London's airports (Heathrow and Gatwick).

At a specific level the locational choice of an office occupier is determined by such diverse factors as:

- tradition
- proximity to markets
- staff availability
- quality of housing
- complementary businesses
- availability of parking
- individual preferences by directors.

A similar pattern of choice may exist amongst companies in the same business, the prime example being the City of London, which is the centre for banking, insurance and finance. Service companies started to select different locations for different activities within their companies from the 1980s onwards. For example, the headquarter offices might be in central London with the support facilities sited on an out-of-town business park. Since the advent of IT and with the introduction of more work–life balance options for employees, many companies have introduced 'hot-desking' (i.e. several employees share a desk or workstation) or 'teleworking' (i.e. employees working from home via computer and telecommunication links with head office). There is debate about how such changes in working patterns and the advances in IT will affect demand for office requirements in the future.

Retail

Retail development takes place within a hierarchy of shopping locations. This hierarchy of shopping locations consists of:

- regional centres
- district centres
- local centres
- superstores/retail warehouses

A particular shopping area will be classified within the hierarchy by reference to its general characteristics and the size of its catchment population. The catchment population is typically calculated by reference to the size of the population living within 10–20 minutes driving-time zones from the centre of the area. A regional shopping centre usually has a catchment population in excess of 90,000 people and a district centre usually has a catchment population of less than 90,000 people. Regional centres are usually large towns and cities, e.g. Bristol. However, there are a few regional shopping centres in out-of-town locations, e.g. Meadowhall in Sheffield. District centres are medium-sized shopping centres that tend to cater for weekly shopping needs. Local centres are usually located within residential areas and cater for the daily needs of local residents. Retail warehouses and superstores/ hypermarkets are located on main roads in the edge-of-town centres with good accessibility from nearby residential areas.

When seeking a site a retail developer will analyse the catchment area where the proposed scheme is to be located. In relation to regional and district centres the catchment will be defined in terms of the drive times from the centre and analysed in relation to both the population size and its characteristics in terms of social and economic groups. An analysis will be made of competing shopping centres and the impact of the scheme on those centres. In carrying out this analysis the developer is assessing the trading potential of the proposed scheme within the hierarchy of shopping centres. In other words, the site is being looked at from the point of view of the retail tenant and the 'shopper' since the success of the scheme will depend on this. This type of analysis will be discussed in Chapter 7.

In the case of retail or bulky goods warehousing, there are many developers who specialise in out-of-town retailing. They develop a good, detailed knowledge of the locational requirements of each retailer through their working relationship with them. Many of the retailers who operate in out-of-town stores, particularly the food retailers such as Tesco in the UK, actually carry out their own developments. The development division of the retail company identifies potential sites while the retail division considers the potential trading position.

A critical factor in the precise location of any proposed town centre retail scheme, whether it is a single shop unit or a major shopping centre, is pedestrian flow. Studies are carried out into pedestrian flow, which is influenced by car parks, bus and railway stations, pedestrian crossings and major stores. It is the precise location of a shop or store that determines its rental value and it is vital the developer gets it right. In town centre locations, shops are either classified as 'prime', 'secondary' or 'tertiary' by virtue of the pedestrian flow and proximity to the major stores.

Since the mid-1980s, retail development activity has been concentrated in out-of-town and edge-of-town locations, with schemes such as Bluewater in Kent becoming the most successful of trading places for retailers in the country. The majority of retail floor space is in town and city centre locations. However, many town centres lost trade to competing out-of-town schemes, and the government responded in 1993 by issuing revised planning guidelines for retail development. Planning Policy Guidance Note 6 (PPG6) (see Chapter 5) places greater emphasis on town centre and edge-of-town locations for future retail development. As a result of the guidelines it became more difficult for developers to obtain planning consent for out-of-town retailing, unless they could provide evidence that the nature of the development did not affect the viability of the town centre.

Industrial

There are various types of industrial property, each of which has different locational characteristics. Industrial property can be categorised into warehousing/distribution, light industrial and general industrial. Broadly speaking, the bulk of factory space in England is located in Yorkshire, the north of England and the Midlands. However, factory development does occur in other regions, sometimes with the assistance of government incentives. The location of factories and heavier industrial space is largely due to tradition and the lower cost of both premises and labour. Often foreign manufacturers and relocating British manufacturers are attracted to the above-mentioned regions by the existence of grants and other incentives.

A warehouse is a building used for storage or distribution and, more recently, retail sales. Warehouses are occupied by retailers, manufacturers and distribution companies and in recent years their requirements have become highly specialised due to technological advances. Accordingly, most warehouse development is carried out on a 'design and build' rather than a speculative basis to suit the individual tenant. Generally, the bulk of warehousing in England is located in the North, the West Midlands, the South East and near London as it is central to all national and regional markets in the UK. Sites suitable for warehousing must be located with good access to the motorway network.

The term light industrial defines processes that involve the manufacture of goods without any environmental impact. There are various types of occupier who require light industrial property. For example, there is the traditional manufacturer, and there are 'high tech' companies who require office, research and development facilities and production facilities. Under current planning legislation (see Chapter 5), office, light industrial and research and development uses are classified as one class of use, and an occupier can change from one use to another without requiring planning permission from the local authority. As such, light industrial buildings are developed to a specification and in a location to suit the type of potential occupier. 'High tech' companies, for example, who combine production with research and development activity, have preferred the south of England as they employ graduates who prefer to live there.

There are general factors that influence the location of industrial premises. Industrial occupiers need to locate in areas:

- close to their markets
- close to supplies of raw material
- with good access to major roads (note that any access via secondary roads needs to be good and uncongested.)

Companies who employ a high proportion of office and research and development staff will often have similar locational requirements to those of office occupiers, i.e. an attractive environment and the availability of quality housing.

Once a developer has established the areas of geographical search, the next stage is to define more specifically the preferred locations within a town or area. This will enable the developer to establish a strategy and brief for the site-finding process. It is important when planning a land acquisition strategy to define the size of target sites. The definition of size might be related to acreage or the investment lot size of the potential development scheme.

Discussion point

What factors do developers take into account when identifying sites for various different types of development (e.g. retail, office or industrial developments)?

2.3 Brownfield and greenfield sites

Since the 1990s there has been an increasing emphasis placed on urban regeneration. The need for increased urban regeneration was one of the key findings of the influential 'Urban Task Force', which was chaired by Lord Rogers of Riverside and reported in 1999 (Rogers, 1999). The report concluded that there was a need to build large quantities of housing for people in the UK and that existing density of housing development was low compared to other cities. Barcelona, for example, has average densities of 400 dwellings per hectare where Islington, London, has around 100–200 per hectare (Syms, 2002). The aim of urban regeneration has been supported by major changes in government policy within the UK to promote the reuse of land, known as 'brownfield'. The development of brownfield land is actively promoted while greenfield development (or previously undeveloped land) is discouraged. A well-known example of brownfield development is the London Docklands. Urban regeneration, or 'urban renaissance' as it is sometimes referred to, is increasingly undertaken in all developed countries in urban centres. Very often brownfield land has a good location, being sited in city centres, which is appealing to property developers. Examples of redevelopment of brownfield land in docklands locations are to be found in cities such as Toronto, Montreal, Melbourne and Vancouver.

However, the redevelopment of previously developed land and buildings is complex and involves risks, for example dealing with contamination. Rogers and Powers (2000), and others, have reported that many of the best brownfield sites in the UK have already been redeveloped and the remaining sites are often poorly located or the costs of cleaning up contamination are prohibitively high. The issue of contaminated land is dealt with in more detail later in this Chapter (see 'Contamination'). It is likely that in some parts of England there are locations where there is an insufficient supply of brownfield land for residential and commercial property development. Syms (2002) argued that the same principles of increasing density of development should be applied to greenfield developments, although not as high as the inner-city development densities, thereby reducing the amount of land required to accommodate the development.

2.4 Initiation

Having researched and defined a strategy for site acquisition, the developer's next step is to seek and identify potential development sites. This can be achieved in a number of ways. However, before we discuss the various methods of site finding, it is important to realise that theory and practice often differ. A developer may have a well thought out and thoroughly researched land acquisition strategy, but achieving that strategy is dependent on numerous factors, with many beyond the control of the developer. This is where the property development process is unique and much depends on the opportunities available. A developer's ability to acquire land is dependent on the availability of land at any particular time. The availability of land is dependent on the state of the market, planning policies and physical factors, and any particular case will also depend on the motives of the particular landowner. The various types of landowners and their motives for owning land were discussed in Chapter 1. The developer, landowner, agent and public sector are the main stakeholders involved in the initiation process. The landowner may take either an active or passive role in the process.

Generally more land is likely to be available when land values are rising rapidly. The availability of land will be influenced by the allocation of land within a local planning authority's 'development plan' and the perceived chances of obtaining planning permission in respect of unallocated areas of land or land allocated for other uses. Although land may be available on the market and is allocated within the development plan for the proposed use, it still might not be suitable for development due to physical factors. For example, a lack of infrastructure such as roads and services might make a development scheme not viable. Also the state of the ground, which might be contaminated or unstable, may prohibit profitable development.

The various ways of initiating the site acquisition process are examined below.

Developer's initiatives

Firstly, there is no substitute for identifying a site by knocking on people's doors. A developer may employ an in-house team, an agent or a planning consultant to find development sites based on the criteria set out in the site acquisition strategy. Many developers, particularly those who specialise in a certain type of development (e.g. retail warehousing) and the large house-builders, employ 'acquisition surveyors' or 'land buyers' and their role is to find and acquire sites in accordance with the company's strategy. The developer needs to acquire an excellent knowledge of the target area and thoroughly research relevant planning policies. Finding sites by this method may incur considerable hard work with no results. Searches can be made by car, foot or by other means to identify potential sites. This done, the next step is to ascertain who owns the land, and there are numerous ways to achieve this, including examining the planning register, contacting the Land Registry, asking local agents or literally knocking on the door.

In England the planning register is a register of all planning applications and permissions in a particular planning authority's area. When a planning application is made, the owner of the land to which the application relates must receive a statutory notice from the applicant. Therefore, an examination of the register will reveal the owner of a piece of land, provided a planning application has been made. However, the details of the landowner may be out of date. Local authorities hold a register of publicly owned vacant and underused land, which is also available for inspection. Since 1990 the Land Registry, the statutory registry of all legal titles to freehold and leasehold land in England and Wales, has been open to the public. The Land Registry maintains a computer-based statutory register of around fifteen million properties in England and Wales. Developers can apply to the Land Registry to establish the name and address of the owner of a property, if it is registered, for a fee.

A developer may also employ a planning consultant to carry out a strategic study of a particular area, which will involve identifying suitable land within the planning context. A strategic study of an area will involve examining the 'development plans', i.e. structure plans/unitary plans and, where they exist, local plans (see Chapter 5) covering that area, a study of the planning register and discussions with the planning officers of the local authority. The study will usually identify sites that have been allocated in the development plan but not yet developed, commenting on their suitability and availability for the proposed use. A report will be made on each site describing its characteristics, planning history and details of the landowner if known. The study will also identify sites that have not been allocated but where there is a good chance of obtaining planning consent by negotiation or on appeal. The best time to undertake this study is when the development plan is in the draft or review stage as there is then a chance to influence the allocation of land by presenting evidence at the public inquiry to include a particular site. Therefore, it is crucial that developers know the timetable of every review and draft publication of the development plans relevant to their area of search. The study should advise the developer which sites should be pursued.

A developer may employ a particular agent or lobby a number of agents to find sites in a particular area. The developer briefs the agent as to the requirements in terms of the nature and size of sites. The agent should have a good knowledge of the area and its planning policies, and often a local agent is employed or approached. The developer will develop contacts with a number of agents, and it is important to develop good relationships to ensure the agents stay loyal. If the agent is retained directly by the developer a fee will then be payable if the latter is successful in acquiring a site identified by the agent. This is typically one per cent of the land price, but it is very often a matter of negotiation dependent on the agent's subsequent involvement in the development, letting and funding of the scheme. The advantage of using an agent is that they become the developer's eyes and ears. Through their knowledge of the area, they know who owns a particular site and its history, as well as being able to anticipate whether a site may come onto the market, for example through their contact with local landowners. With occupied buildings they may know when leases will expire and thus when possible redevelopment opportunities might arise.

The advantage of early identification of sites is that the developer has the opportunity to negotiate directly with the landowner and secure the site before it goes on the market. A developer's ability to acquire sites off the market will depend not only on negotiating abilities but the actual state of the market. When the market is booming and land values are rising rapidly, then the landowner may be strongly advised by agents to put the site on the open market. A negotiated deal may not be possible if the landowner is a local authority, as they are publicly accountable and the District Valuer must be satisfied that the right price has been achieved.

Developers may also identify sites in less obvious ways. They may acquire a whole company as a means of securing a site or an entire portfolio of properties that fit in with their acquisition strategy, e.g. some development companies have acquired retail chains as a means of securing 'prime' high street shop units. In these situations the developer will retain the ownership of the property assets and either sell the operating part of the business or wind the business up. Developers also may acquire individual properties or entire portfolios through direct approaches to other developers or property investment companies.

Agent's introductions

Although agents may be retained by a developer to find sites, they will also take the initiative and introduce opportunities to developers directly. The opportunity may be a site already on the market or a site that is likely to come on to the market shortly. If the introduction to the developer ends in a successful acquisition of the site, then the agent will expect a fee from the developer, unless they are retained and instructed by the landowner. The fee is typically one per cent of the land price but it may be a negotiated fee dependent on the agent's subsequent involvement with the scheme through letting and/or funding instructions.

An agent will introduce a site to the developer who is most likely to be successful in acquiring that particular site. Some agents remain loyal to a developer because that particular developer is an established client and they have a good working relationship. The agent will look at the experience of the development companies and their financial status when making their decision about who to introduce a particular site to. The most likely candidates are those developers who are active in the particular market at the time, whether it is, for example, out-of-town retailing or industrial schemes, and have a history of successfully bidding for sites.

A development company, depending on its size and financial status, may receive introductions on a daily basis when market conditions are favourable. It is important to set up a register of sites that have been introduced to the company because it is likely that different agents will introduce the same scheme to different people within the same company. It is important to avoid duplication of agents, otherwise two acquisition fees may be payable.

When introducing a site, the agent should provide enough detail to enable an initial decision to be made by the developer as to whether or not to pursue the opportunity. Ideally, the particulars should include a site plan, location plan, planning details, and details of the asking price and terms. The introducing agent is responsible for assisting the developers throughout the acquisition. The agent should advise on the local property market and rental values to help the developer in the evaluation process. Information on existing and proposed schemes of a similar nature is also vital. The agent will often negotiate the land price on behalf of the developer.

This method of site finding is a two-way process. The developer must establish and maintain a good relationship and regular contact with local and national agents. It is important to provide those agents with details of site requirements to avoid a situation where site opportunities are continually rejected and the agent gives up and goes to a rival developer. Agents should provide a good service to their developer clients as there may be chances that letting and funding instructions will flow from the initial introduction. Other property professionals such as solicitors, planning consultants, architects and quantity surveyors may introduce opportunities to developers. There is some truth in the saying that 'it is who you know not what you know' when related to property development.

Landowner initiatives

A landowner may take an active role in initiating the development process by a decision to sell their land or enter into partnership with a developer. An obvious source for identifying development sites for sale is advertisements, whether in the media, internet, on a site board or via direct mail. All the property publications e.g. *Estates Gazette*, *Estates Times* and *Property Weekly*, carry advertisements each week for sites and development opportunities.

A developer may receive particulars of a site for sale direct from a landowner or their agent. Where a site is advertised in the open market the developer will be competing for the site. The degree of competition depends on how the site is offered to the market. There are various methods including:

- Informal tender
- Formal tender
- Competitions involving one or several short listings
- Auctions

The method of disposal will depend on market conditions and the motives of the landowner. The developer may be in competition with any number of others or there may be a selective list of bidders. The various methods used by landowners to secure purchasers or partners are detailed below.

Informal tenders and invitations to offer

Informal tenders or invitations to offer involve inviting interested parties to submit their highest and best bids within a timescale. This usually involves all parties who have expressed an interest in the site and the invitation may include an indication of the minimum price acceptable. For example, it might state that offers of over £1,000,000 are invited and indicate what conditions attached to the bid may or may not be acceptable to the landowner.

The important point from the developer's perspective is that the bid made is subject to any necessary conditions. After a bid has been accepted by the landowner, the developer has the ability to renegotiate the price if there is some justification to do so before contracts are exchanged, though there is always a risk that the landowner may not accept a revised price and may go to another party who made a bid. Developers prefer informal to formal tenders as they allow bids to be made on the developer's own terms. However, the more conditions a developer attaches to a bid the less likely it is that the bid will be acceptable, even if it is the highest received. The landowner generally accepts the highest bid unless the conditions are unacceptable or the developer's financial standing is questionable. After receiving the bids, the landowner may negotiate with several of the parties before making a decision in an attempt to vary conditions or the level of the bids.

Formal tender

A formal tender binds both parties to the terms and conditions set out in the tender documentation subject only to contract. It involves an invitation to interested parties to submit their highest and best bids by a deadline. The invitation sets out the conditions applicable. The document will usually state that the landowner is not bound to accept the highest bid. This method may or may not involve a selection of the interested parties. Developers, generally, do not prefer formal tenders as it reduces their flexibility and increases their risk. The exception would be a situation where all the possible unknowns had been eliminated – for example where a detailed, acceptable planning consent was present, a full ground and site survey had taken place and the site was being sold with full vacant possession.

Competitions

Competitions are used by landowners when financial considerations are not the only criteria for disposal of the site. Therefore, competitions are mainly used by local authorities and other public bodies seeking to select a developer to implement a major scheme. They are used more informally by other landowners seeking development partners – for example, a landowner may want to obtain planning permission before disposing of the land and therefore the developer may be selected on the basis of planning expertise. Alternatively, the landowner may not wish to dispose of the land and will seek a developer to project manage the scheme in return for a profit share.

As the majority of competitions involve local authorities and other public bodies, this discussion is confined to public authority competitions. Local authorities and public bodies invite competitive bids on a tender basis, whether formal or informal, and the bids will normally be judged on a financial and/or a design basis. As a first step, the authority will usually advertise their intention to set up a competition and invite expressions of interest. Alternatively, the authority may choose the developers to enter the competition.

If the former method is adopted, usually developers will be invited to express their interest in becoming involved. They will generally be asked to provide details of their relevant experience and track record, financial status (most often a copy of their company report and accounts), the professional team if appointed and any other information that is relevant. For example, a developer may own adjoining land to the competition site or may have been involved with the subject site for some considerable time.

The authority will assess the expressions of interest and compile a shortlist of developers to enter the competition. This may or may not be the final selection process, and bids may be invited from those shortlisted in order to compile a final shortlist. The number of selection processes will depend largely on the number of interested parties and the complexity of the competition. If the authority is asking developers to submit both financial and design bids, and the design being asked for is fairly detailed, then the number of developers shortlisted for the final process should be no more than about three. Many competitions involve developers risking the large sums of money they spend, and in these circumstances long shortlists are not favoured.

It is important that a development brief is prepared to provide guidelines for the competition, including a statement of the basis of the competition and on what criteria the developer will be selected. The development brief will set out the requirements of the authority with regard to such matters as total floor space, pedestrian and vehicular access, car parking provision, landscaping and any facilities that the authority considers desirable in planning terms. The authority may include a sketch layout or outline sketch drawings illustrating the development required, but in the majority of cases it is the developer's responsibility to suggest design solutions. The brief should state how flexible the authority will be in assessing whether the bid meets its requirements. It is vital that a developer finds out whether they will be penalised for not strictly adhering to the brief. Generally developers who follow the guidelines in the brief will be looked upon favourably, unless a developer proposes a better solution to that outlined in the brief. It may be that through their ability and expertise a developer may produce a higher financial bid by proposing more square footage (or square metres) than that envisaged in the brief whilst still producing a good and sensitive design. Every competition is different and it pays the developer to study the development brief in depth and look at all possible angles that can be used to advantage.

Developers find competitions the least attractive method of acquiring development sites as the process demands time and expense in the preparation of drawings and financial bids, which are aborted if the developer loses.

Auctions

Some development sites are sold at auction. Many of these are unusual sites. For example, Railtrack uses auctions to dispose of disused railway embankments and land with no, or limited, access. Auctions may sell property investment opportunities, such as when the leases are due to expire in the next 5 years and there is obvious redevelopment potential. It pays a developer regularly to look through auction catalogues for development opportunities.

At auction the highest bid secures the site providing that the reserve price has been met or exceeded. The landowner will instruct the auctioneer of the reserve price, which is the lowest price acceptable. If the reserve price is not reached through the bidding, then that particular lot is withdrawn. The auction will set out the standard conditions of sale and special conditions of sale relating to each particular lot. Once a bid is accepted the successful bidder exchanges contracts at that point by handing over the deposit and details of their solicitor. Thus, if a developer acquires a site at auction a thorough evaluation and all other preparatory work needs to be carried out beforehand. It is possible that the lot may be acquired prior to auction by direct negotiation with the landowner.

Competitions and tenders are the most common methods of disposal used by landowners when market conditions are good. However, developers usually prefer to obtain a site off the market, thereby avoiding competition. If a developer enters a number of competitions and tender situations then they could all be lost, or possibly all or some could be won. There is no certainty and the developer may become very frustrated, wasting considerable time and money in the process. Success is based on the developer's ability to judge the right opportunities to pursue and the right level at which to submit a financial bid. However, in many instances it may be a case of luck and being in the right place at the right time. The site acquisition process is very competitive and it must be realised that even the best thought out acquisition strategy may not be achieved in the way, and in the timescale, first envisaged.

Discussion point

What options does the developer have to seek and identify potential sites for development?

Local authority initiatives

Due to central government policy the role of the public sector in the development process has become more limited. The emphasis of central government policy is for local authorities to do the minimum necessary to enable the private sector to participate in development. However, local authorities still have an important role to play in initiating the development process. Their main role is that of planner through the planning system. They may facilitate development by promoting or participating in development opportunities themselves.

Local authorities are restricted by the scope of their legal powers, the availability of finance and the need for public accountability. Some local authorities are more active than others depending on the political party in overall control of the authority and whether they wish to actively encourage development within their area. We shall now examine the various methods by which local authorities influence the availability of land for development.

Planning allocation

The allocation of land within a local planning authority's development plan establishes the framework for the permitted use of land and hence establishes its potential value for the purposes of development. In formulating planning policies in the development plan a local authority has to balance the demands of developers against the wider long-term interest of the local community. The way in which local authorities allocate land through the framework of development plans and how developers can participate in this process is examined in Chapter 5.

Developers will examine the development plan relevant to the areas identified in their search for sites and, therefore, the local authority in their role as planner can influence the availability of a particular site by allocating a specific use to it in the development plan. However, allocating a site in a development plan does not make it available for development. The developer and landowner must be able to agree terms and the site must be suitable in physical terms for the proposed use. Even if it is available it may not be developed because the location does not meet the requirements of occupiers. If the developer and or landowner disagree with a particular allocation within a development plan, in preference to their own site, then they can present their case to a planning inspector at the public inquiry into the development plan.

Land assembly and economic development

Local authorities may make land available for development by assembling development sites for disposal, which may involve their statutory compulsory purchase powers. Their ability to take on this enabling role obviously depends on the amount of land under their control and their attitude towards encouraging development. Some authorities in areas of economic decline and high unemployment actively encourage private sector investment. For example, Bradford City Council planning department works with developers to bring forward sites using their land acquisition and development powers to deal with physical constraints on development. Activity in economically prosperous authorities may be restricted to involvement in prestigious sites, such as town centre shopping schemes. However, positive participation by local authorities in making land available is not just limited to land assembly, whether by agreement or compulsion, but may include site reclamation, the provision of buildings, the provision of infrastructure/services, the relocation of tenants and general promotion of their area as a viable business location. Any or all of these activities tend to be described by the general term of 'economic development'.

The need for a local authority to become involved in 'economic develop-ment' depends on the initiatives taken by the private sector and whether market forces alone meet the expectations of the local authority for the development of their area through the creation of employment opportunities. However, the extent to which local authorities can undertake 'economic development' is restricted by government policy in relation to local authority finance. For example, Part 3 of the Local Government and Housing Act 1989 provides the legal power to participate in economic development; however, local authorities are under no duty to provide services to assist economic development. Under the provisions of the same Act, local authorities can only use half of their capital receipts (i.e. proceeds from the sale of land or buildings) for new capital investment. The remaining half has to be used to redeem debts or has to be set aside to meet future capital commitments or used as a substitute for future borrowing. A local authority's ability to raise money through capital receipts is taken into account when the government sets their credit approval limit, i.e. the extent to which they can borrow money. The definition of capital receipt extends to the receipt of rent (e.g. occupational rents and ground rents) and the receipt of reduced rent in lieu of some benefit (which must be fully valued in monetary terms). In addition, temporary financing by local authorities such as the acquisition of land pending disposal to a developer will count against their credit approval limit if the period between acquisition and disposal runs over a year. This is not unusual in the property development process.

Some local authorities, especially in the inner cities or in areas of high unemployment, receive additional funding from the government. The various grants and funding available are examined in 'Funding and grants' later in this chapter. However, access to government assistance is becoming increasingly competitive and local authorities have been forced to explore more innovative ways to achieve economic development aims, such as PPPs. We shall examine partnerships with the private sector in 'Partnerships with developers' below.

When a local authority is involved in land assembly it can benefit the private sector developer but it will often lengthen the development process. A developer may require the support of the local authority when a site identified by a developer is owned by different landowners – for example in the case of a town centre location. The local authority may have allocated the site for complete redevelopment in the relevant development plan and thus is willing to work with the developer to realise their planning aims. In such a situation the developer may experience difficulties in negotiating reasonable land values with the various landowners as they effectively hold the developer to 'ransom' by demanding unrealistic prices, because their landholding is essential to the proposed development. On the other hand, a landowner may be unwilling to sell their landholding as their motivation for occupation is their business. In addition, a particular development site might be landlocked with access under the control of a landowner who is seeking a price well above the market value because of their advantageous position. A local authority who assists with land assembly in this way will try to reach agreement by negotiation. If this is unsuccessful then they may make a compulsory purchase order under Section 226(1) of the Town and Country Planning Act 1990. Local authorities, subject to the authorisation of the Secretary of State, have power to acquire any land that is in their area and:

- which is suitable for and is required in order to secure the carrying out of development, redevelopment and improvement;
- which is necessary to acquire to achieve proper planning of an area in which the land is situated.

The process of compulsory purchase has been criticised as lengthy, drawn-out and antiquated. It involves the local authority agreeing to compensation values with all the interests (i.e. any interest in the land including weekly tenancies) affected by the compulsory purchase order (CPO). Consequently it was little used since its heyday in the 1960s and 1970s; however, the Planning and Compulsory Purchase Act 2004 is meant to address these issues and speed up the process. In the case of an urban redevelopment the number of interested parties can be in the hundreds, and even in the thousands for new road or railway lines through urban areas. Notice must be served on all interests involved and details of the scheme publicised. Every CPO involves a public inquiry, at which all interested parties may present evidence to an inspector appointed by the Secretary of State. Then the Secretary of State, if satisfied with the case presented by the local authority, will confirm the order, which will provide the local authority with the necessary powers to acquire the land. Every claim for compensation is then agreed between the valuers acting for the local authority and the affected party based on the rules of compensation set out in the Land Compensation Act 1961. The compensation payable for the compulsory acquisition of land is based on the principle that the owner should be paid neither less nor more than his loss. Thus it represents the value of the land to the owner, which is regarded as consisting of:

- the amount which the land might be expected to realise if sold on the open market by a willing seller (open market value);

- compensation for severance and/or injurious affection;
- compensation for disturbance and other losses not directly based on the value of the land.

Where the property is used for a purpose for which there is no general demand or market (e.g. a church) and the owner intends to reinstate elsewhere, he may be awarded compensation on the basis of the reasonable cost of equivalent reinstatement.

Local authorities will appoint an independent valuer or the district valuer to act on their behalf. In the event of disagreement between the parties the matter is referred to the Lands Tribunal for resolution. Further appeal on a point of law may be made to the Court of Appeal.

It is not possible in this book to explore further the details of the compulsory purchase procedure. However, it is worth noting that the basis of compensation has been criticised for inadequately compensating affected parties for the losses incurred. Another problem caused by compulsory purchase is blight, which often affects the areas allocated for redevelopment, where the authority has indicated an intention to use compulsory purchase powers. Owner occupiers can serve a blight notice to force the local authority to acquire their land if they have been unable to sell their interest or achieve a price equating to the value of the land without blight. However, this option is only open to commercial owner–occupiers whose property has a rateable value below a specified limit. Note there are no such restrictions on residential occupiers.

In return for assisting a developer with site assembly the local authority may negotiate the provision of social facilities or participate in the financial rewards of the eventual development scheme. However, it can be known for some local authorities in their role as planners to use the threat of their compulsory purchase powers in a negative manner to achieve some material benefit in the form of 'planning obligation' (see Chapter 5) or amendments to planning applications.

When local authorities dispose of land to developers they usually produce a development brief that outlines how they would like to see the site developed. It is important that the brief is flexible and not too detailed to allow the developer freedom to react to prevailing market conditions. Where compulsory purchase powers are used the land assembly process may have taken several years, by which time market conditions may have changed significantly. It is important from a developer's point of view that sites are disposed of on as clean a basis as possible. In other words any problems or constraints that exist with regard to the legal title, services, planning and access should be tackled from the start.

Infrastructure

The provision of supporting infrastructure is critical to the site acquisition process and it is an area in which local authorities play an important role. Infrastructure is a term used to describe all the services that are necessary to support development, i.e. roads, sewers, open space, schools.

As discussed above (see 'Site identification'), the existence or proposed provision of roads is vital in assessing locations for property development. Whilst proposals for a new road will generate pressure for development along its route, new development will create new traffic pressures on the existing road network. As infrastructure is so critical to the viability of a particular development scheme it directly influences land values. If the necessary infrastructure does not exist to support a development then a developer will take account of the cost of its provision in the evaluation of the land value (see Chapter 3 for further details on the evaluation process). Infrastructure has a great impact on the sustainability of a development (see Chapter 9 for further details).

Local authorities are largely responsible for deciding the level of infrastructure required and securing its provision, and they have to determine who is responsible for its cost. Due to government control on spending, local authorities often negotiate agreements with developers (known as planning obligations – see Chapter 5) to secure the provision of new infrastructure, if it is needed to support the development, e.g. the provision of a roundabout to link the development scheme with the existing road network or the provision of public open space. The assessment of future infrastructure requirements at a strategic level is the responsibility of the county council or the unitary authority (see Chapter 5) as part of the process of preparing development plans. In most cases the county council or unitary authority is the highway authority for their area, although they may delegate some responsibility to district councils. The highway authority will be consulted on all planning applications to establish whether a development can be supported by the existing road infrastructure. The Department for Transport (Highways Agency) is responsible for motorways and trunk roads. Water companies are responsible for the provision of sewers and the water supply, the cost of which is agreed directly with a developer where it relates to a particular development.

Many authorities adopt a more active approach to the provision of infrastructure; they recognise that new roads open up land for development. Land is often assembled in conjunction with the new road so that the authority can benefit from enhanced land values by packaging sites for disposal to the private sector. There is much debate about the pressure for development caused by new roads, particularly in prosperous and environmentally sensitive areas, and how the cost of provision should be allocated. At the moment developers may be required to enter into Section 106 or Section 278 agreements (under the provisions contained in Sections 106 and 278 of the Town and Country Planning Act 1990) with the relevant highway authority or the Department for Transport, to secure a financial contribution to pay for improvements to existing roads required to accommodate traffic from a particular development proposal. Planning guidelines relating to traffic exist in the form of Planning Policy Guidance Note 13 (PPG13), which affects the preparation of policies relating to the location of traffic-generating developments in future development plans. The government wants to reduce car travel by influencing the location of development schemes relative to the existing road network and public transport. They wish to encourage development that is accessible via alternative forms of transport such as walking, cycling and public services. Overall the aim is to encourage local authorities to co-ordinate land use policies and transport infrastructure to achieve a reduction in travel needs (for further details on PPG13 see Chapter 5). In addition the government review major road building projects, since evidence suggests that new roads create more traffic congestion.

Many local authorities are trying to promote public transportation in their areas, such as light railway/tram systems, to ease traffic congestion – for example, the 'Supertram' in Manchester. In order to secure the necessary public funding for such transportation systems, government regulations stipulate that private sector contributions have to be secured in advance. Often developers and landowners with sites that will benefit from the system proposed will be approached. However, there is a limit to how much developers can contribute as any contribution will be deducted from the land value the developer will be able to pay. The same applies to contributions made by developers for road improvements, as discussed above.

Many authorities have adopted a positive approach to improving the infrastructure of their town centres, in the face of competition from out-of-town retailing, with schemes to improve the overall environment and parking provision (which may conflict with the provisions of PPG13). Town centre managers are often appointed to propose and implement improvement schemes in partnership with the private sector. Retailers, such as Boots, have contributed towards the salaries of town centre managers. However, local authority resources are limited as revealed in the Department of the Environment's report in 1994, *Vital and Viable Town Centres* (DSDNI 2007).

Partnerships with developers

'Site investigation' below examines some methods adopted by local authorities in disposing of their landholdings. Many authorities retain a legal interest in development schemes by granting a long leasehold interest to the developer, receiving a ground rent usually linked to the long-term success of the scheme. Alternatively, they may retain an interest until the scheme is complete, granting a building licence for a nominal premium to a developer to enter onto the land and complete the scheme. Under this arrangement the authority sells the freehold to the developer on completion and, therefore, shares in any uplift in values. Although the arrangements above are presented to the public as partnerships, they are not true partnerships as the private sector bears the majority of the risk whilst sharing some of the rewards.

Development schemes involving local authorities can involve lengthy and costly competitions to choose a development partner. The size of such schemes involves a significant risk for the private sector, with substantial sums of money expended before the development funding is secured. The legal agreement between local authority and developer often takes a long time to negotiate and is usually subject to the securing of finance by the developer. At times developers withdraw from such schemes due to lack of funding or because the initial evaluation of the scheme has altered due to the passing of time. There is often a conflict of motivations with the local authority's social objective and the private sector's profit objective. The local authority also has a number of roles in the development process that may conflict with each other, such as planner and landowner. A local authority is run by democratic processes that are lengthy and inflexible compared to the faster decision making of the private sector. It often takes time for developers to gain confidence amongst the elected members of the local authority, only for changes to occur in the personnel and the political party in overall control.

Many authorities, as an alternative to the above, have entered into joint ventures with developers via companies limited by shares or guarantees. Also some have formed wholly owned subsidiaries, for example enterprise boards to carry out economic development initiatives. However, Part 4 of the Local Government and Housing Act 1989 regulates and restricts the power of local authorities to set up and participate in companies. This Act limits the amount of interest a local authority can have in a company without the company being treated for accounting and expenditure purposes as an extension of the local authority. Central government wants to ensure that local authorities remain accountable to the public and stops them using joint companies as a means to avoid capital expenditure restrictions. Developers normally prefer joint venture arrangements as they are familiar to them and allow quicker decision making. In addition, they allow more flexibility in securing funding for the scheme as the developer passes some risk to the local authority who shares in any decrease in value of the scheme.

In the mid-1990s central government relaxed Treasury rules concerning financial arrangements made with the private sector by the public agencies responsible for urban regeneration (UDCs and English Partnerships). In the last decade or so, Public Private Partnerships (PPPs) have become another method by which developments can be procured involving a partnership between the private sector and local authorities (see Chapter 6 for further details).

Discussion point

In what ways do local authorities assist the development process in respect of land for development?

2.5 Site investigation

We have examined various methods by which developers identify and secure sites. It is now important to examine the different investigations a developer should make before site acquisition, as these investigations affect the conditions the developer will seek to impose in the contract to acquire the site. Although landowners, particularly local authorities, will provide as much information as possible, it is up to developers to satisfy themselves before entering into a commitment to acquire a site. The following investigations are vital when a developer is acquiring a site, as for example the investigations may show that the proposed scheme is no longer viable due to the physical state of the ground and the cost of remedying the problems.

Site survey

Firstly, a site survey needs to be undertaken by qualified land surveyors to establish the extent of the site and whether the boundaries agree with those shown in the legal title deed. This is important where a site is assembled by bringing together various parcels of land in various ownerships. In this instance the survey needs to establish that all the boundaries of the various parcels dovetail together and that the whole of the site is in fact being acquired. It would be disastrous if the developer discovered midway through a scheme that a small but vital part of the site had not been acquired. The developer would then have to negotiate from a very weak position with that landowner, effectively being held to ransom. The site survey also establishes the contours and levels of the site. It will be necessary to undertake a structural survey if any existing buildings on the site are to be retained in a scheme.

Furthermore, the legal search of the title deeds needs to establish who is responsible for the maintenance of the boundaries. In addition, the access arrangements to the site need to be checked to ensure that the site boundary abuts the public highway. If a public highway exists the solicitor needs to check whether it has been adopted by the local authority and is maintainable at their expense. If access to the site is via a private road, then the ownership and rights over that road need to be established.

Ground investigation

Unless reliable information already exists regarding the state of the ground then a ground investigation needs to be undertaken by appropriate specialists. Ground investigations can vary in cost and extent depending on the proposed scheme and information already known. An investigation will usually take the form of a series of boreholes and trial pits taken at strategic locations on the site. Samples taken from the boreholes and trial pits need to be analysed in a laboratory to establish the nature of the soil, substrata and water table, together with the existence of any contamination. The issue of contamination will be examined further below.

The results of the investigation will be given to the structural engineer, architect and quantity surveyor. They will need to analyse the results to establish whether any remedial work is necessary to improve the ground conditions or whether any piled foundations are required, e.g. where the ground is made up with fill material. Both circumstances will have an impact on the cost of the development scheme, which may affect the overall viability.

Contamination

Property development is largely about risk and managing risk, as such the existence of any contamination on a site is an issue that developers cannot ignore due to legislation and litigation. Contaminated land is defined in Part IIA of the Environmental Protection Act 1990 (EPA) Section 143 as 'land which is being or has been put to contaminative use' (and was brought into being by way of section 57 of the Environment Act 1995). The statutory definition of contaminated land (EPA 1990, s 78A (2)) is:

any land which appears to the local authority in whose area it is situated to be in such a condition, by reason of substances in, on or under the land, that –

- significant harm is being caused or there is a significant possibility of such harm being caused; or
- pollution of controlled waters is being, or is likely to be, caused.

The important inference here is that 'significant harm' is resulting or is likely to result from the contamination, and not that there is possible harm from pollutants on the site. Clearly there is also much debate on the point at which 'significant harm' results as there is no definitive point where this occurs because each land parcel is unique in terms of geology, extent of pollution and so forth. There are three key variables that impact on the amount of harm that can result and these are:

1. the type of contaminant
2. the pathway
3. the receptor.

Different types of contaminant lead to different types of harm. The extent of harm is also dependent on the pathway or way in which the contaminant is able to migrate or move around the site. Therefore, the geology of the site has an important effect, for example soils that are dense or solid allow less opportunity for pollutants to move around. Finally, receptors vary from being animal, plant or human and clearly different pollutants and concentrations will have different effects depending on the receptor. In addition, some pollutants, for example the presence of zinc, can make plant growth verdant and abundant, whereas the effect on human health is adverse. Environmental consultants will examine the relationship between these three variables to establish whether 'significant harm' is likely.

The emphasis of government policy on environmental legislation is based on the principle that 'the polluter pays', following the example set by European Union (EU) legislation. Contamination is typically caused by a previous occupier/s use of the land; however, with many sites the use of contaminants is unrecorded, which makes it incumbent on the developer to undertake a thorough investigation, especially of brownfield sites. Occasionally contaminants will migrate onto land from other sites and thus pollute the land, although such an occurrence can cause complex issues to arise in terms of getting the polluter to pay costs of remediation. It is worth noting that the types of uses that have lead to contamination include:

- manufacturing and industry sites
  - asbestos works and buildings containing asbestos
  - dry cleaners
  - food processing plants
  - metal mines
  - brickworks and potteries
  - chemical works
  - steel works
  - plating works
  - munitions factories and test sites
  - paint works
  - paper and pulping works
  - printing works
  - tanneries
  - textile mills
- infrastructure sites
  - cemeteries
  - docks canals and shipyards
  - quarries
  - railway land
  - sewerage works
- industrial storage sites
  - landfill sites
  - military airfields
  - oil and petroleum refineries and storage sites
  - scrapyards
- power generation
  - gas works
  - nuclear power stations

Since 2000 the emphasis has been placed on the actual condition of the land itself and whether this would be likely to cause 'significant harm'. Due to the potential for land blight, which arose from earlier policies towards contaminated land, the government in the UK introduced the concept of the 'suitable for use standard' with regards to remediation. Hitherto a policy of a total clean-up or gold standard had been adopted towards remediation; however, the costs of remediating land to such a high standard meant that some development proposals were no longer viable. The concept means that any contamination is considered with regards to the use to which the land will be put. For example, a site with a proposal for industrial units would not have such high remediation standards as a proposal for residential development on the same site.

The cost of ground investigation is much higher when contamination exists, representing a potentially abortive up-front cost for the developer. As much information should be obtained on the site's history of uses before any ground investigation is started. This is achieved by looking at ordnance survey maps and local authority records, and looking for any other likely source of information, such as previous owners. In areas of the country where contamination is widespread, the local authority may have already compiled records of contaminated land, e.g. in Sandwell Metropolitan Borough Council where up to 80 per cent of sites are contaminated. Local authorities also hold information on waste disposal sites in their area. However, information obtained from records may be limited and will always need to be double-checked.

The ground investigation will involve taking soil samples down to the water table level, and extensive surveys of all underground and surrounding surface water due to the risk of contaminants seeping through water. The results of the ground investigation will enable an assessment to be made of the extent and cost of remedial measures. There are many different ways of treating contamination including simply removing it, treating it in-situ or keeping containment either under a blanket of clean earth or a capping. Ongoing measures may be required once the development is complete, such as venting the methane gases to the surface. If contamination is limited to one area of the site it may be possible to design around the problem, for example by locating a car park there. If ground has to be filled with imported material as part of the process then, depending on the standard of treatment required, deep-piled foundations may be required. However, it is possible for developers to find, once on site and in the construction phase, that given the selective nature of the ground investigation survey unfortunately the contamination may be more widespread than first reported. Such a situation can impact on the financial viability of a project and represents yet another risk the developer has to manage and assess. Broadly speaking, when a developer is faced with a contaminated site then the remedial measures often preclude all but the higher value land uses such as retail warehousing. Note, however, that grant assistance is available in some areas and is discussed in 'Funding and grants'.

In view of the increasing concerns about issues of contaminated land, the move towards urban regeneration, the increased use of brownfield sites, and the debate about who should be financially responsible, then it is worth developers spending time to thoroughly investigate its existence. The appropriate professionals should undertake a full environmental audit, which can then be presented to potential purchasers, financiers and tenants.

Services

The site survey should establish the existence of services including water, gas, electricity and drainage. All the utility companies should be contacted to confirm that the services surveyed correspond with each of the companies' records. In addition, the capacity and capability of the existing services to meet the needs of the proposed development should be ascertained. If the existing services are inadequate, then the developer will need to negotiate with the company concerned to establish the cost of upgrading or providing new services, e.g. a new electricity substation. Where work needs to be carried out by either an electricity or gas company then the developer will probably be charged the full cost of the work, although a partial rebate is usually available once the development is occupied and the company is receiving a minimum level of income. The route of a particular service may need to be diverted to allow the proposed development to take place. The cost of diversion and the time it is likely to take to complete should be established at the earliest possible stage. The legal search of the title deeds will reveal if any adjoining occupiers have rights to connect to or enjoy services crossing the site. The developer may need to renegotiate the benefits of these rights if they affect the development scheme.

Legal title

Solicitors will be appointed by the developer to deduce the legal title to be acquired and to carry out all the necessary enquiries and searches before contracts are entered into with the landowner. The developer's solicitor will apply to the Land Registry to examine the official register of the title.

If the land to be acquired is leasehold the register will reveal brief particulars of the lease and the date it was entered into. The developer will need to establish the length of the lease, the pattern of the rent reviews and the main provisions of the lease. Such provisions need to be checked to ensure the terms are acceptable to the provider of development finance. The solicitor needs to establish that the land will be acquired with vacant possession and that there are no unknown tenancies, licences or unauthorised occupancies. The fact that a site or building is unoccupied does not necessarily mean that no legal rights of occupancy exist. The register will also show whether there are any conditions or restrictions affecting the rights of the landowner to sell the land. In addition, all rights and interests adversely affecting the title will be established, such as restrictive covenants, easements, mortgages and registered leases.

The existence of an easement could fundamentally affect a development scheme. An easement may be either positive (e.g. a public or private right of way) or negative (e.g. a right of light for the benefit of an adjoining property). If they exist to the detriment of the proposed scheme then the developer may be able to negotiate their removal or modification to allow the scheme to proceed. Rights of light might affect the proposed position of the scheme and the amount of floor space. If a party wall exists then it will be necessary to agree a schedule of condition with the adjoining property or make a party wall award of compensation. A chartered surveyor with specialist knowledge on party wall matters may need to be appointed by the developer.

Any restrictive covenants may adversely affect a development scheme, e.g. a covenant prohibiting a particular use of a site. Often it is difficult to establish who has the benefit of a covenant as the covenant might have been entered into some considerable time ago. If the beneficiary can be found then the developer may be able to negotiate the removal of the restriction. If not, then the developer can apply to the Lands Tribunal for its discharge, which at times can be a lengthy process. Alternatively, the developer can take out an insurance policy to protect against the beneficiary enforcing it. The insurance cover will compensate against the loss in value caused by any successful enforcement action.

A solicitor will carry out a search of the local land charges register maintained by the local authority. This will reveal the existence of any planning permissions or whether any building or site is listed as a building of special architectural or historic interest. Enquiries will also be made of the local authority to establish whether the road providing access to the site is adopted and maintained at public expense. The existence of any proposed road improvement schemes might affect the site, e.g. a strip may be protected at the front of the site for road widening purposes. Enquiries will also be made of the landowner (vendor) and will include standard questions on matters such as boundaries and services. Enquiries will also reveal the existence of any overriding interests (i.e. rights and interests that do not appear on the register of the title) or adverse rights (i.e. rights of occupiers of the land). Solicitors may also make additional enquiries of the vendor that are particular to the land being acquired.

The developer must aim to acquire the freehold or leasehold title of the development site free from as many encumbrances as possible by renegotiating or removing the restrictions and easements. Financiers, particularly the financial institutions, prefer to acquire their legal interest with the minimum of restrictions that might adversely affect the value of their investment in the future. The developer has to be able to 'sell' the title to purchasers and tenants as quickly as possible without complications.

Finance

Finally, no prudent developer (unless there are sufficient internal cash resources) would consider entering into a commitment to acquire a site without first having secured the necessary finance or development partner to at least cover the cost of acquisition, including interest on the acquisition cost, while the site is held pending development. The developer should aim to ensure that the financial arrangements are completed to coincide with the acquisition of the site. If no financial arrangements are in place then the developer must be satisfied that either the finance will be secured or that the site can be sold on the open market if no funding is forthcoming. The developer must ensure that all investigations have been carried out thoroughly so that any financier or partner has a full and complete picture of the site. Every area of doubt must be removed if at all possible. The developer should obtain specialist advice on the implications of VAT on any site acquisition. The effect of VAT in relation to the evaluation process is described in Chapter 3.

Discussion point

When evaluating a particular site what factors do developers have to consider?

2.6 Site acquisition

The findings of all the above investigations should be reflected in the site acquisition arrangements. The degree to which developers reduce the risk inherent in the property development process depends to some extent on the type of transaction agreed at the site acquisition stage. Prudent developers will always try to reduce risk to a minimum and the site acquisition arrangements are important in this respect. Ideally, no acquisition will be made until all the relevant detailed information has been obtained and all problems resolved. In practice, however, it is virtually impossible to remove every element of uncertainty. The degree to which the developer can reduce risk to the site acquisition stage is dependent on:

- the landowner's method of disposal;
- the amount of competition;
- the tenure

It is possible to pass some of the risks onto the landowner, but this will largely depend on the developer's negotiating abilities.

The majority of site acquisitions are on a straightforward freehold basis. The freehold title transfers from the vendor to the developer once contracts have been completed, and from that point onwards the developer owns the entire risk. The developer reduces the risk inherent in the transaction through negotiation of the contract terms – for example, the contract can be conditioned and payments can be phased or delayed. For instance, if there is no planning consent in existence then the developer should negotiate that the contract is subject to a 'satisfactory planning consent' being obtained. The vendor, if such a condition is acceptable, will try to ensure that the term 'satisfactory planning consent' is clearly defined. The developer may obtain a planning consent that does not reflect the optimum value of the site but satisfies the condition in the contract and then, at a later stage, obtain a better planning consent. It is not uncommon for 'top-up' arrangements to be made whereby the vendor benefits from any improvement created by planning consents obtained by the developer. Developers will carefully weigh the degree of uncertainty in relation to planning and it will be a matter of judgement as to whether the risk involved is acceptable. If the vendor is undertaking to sell the site with vacant possession then the contract should be conditional upon this, since there could be a delay in the occupants of a building leaving.

Whilst the normal period between signing a contract to purchase a site and the completion is 28 days, the developer may negotiate a delayed completion, e.g. 6 months. Delays in the development process cost money, so the developer should ensure that any potential problems revealed by investigations are dealt with before contracts are completed, or the time needed to resolve them is reflected in the evaluation and hence the price paid for the land. As an alternative, especially if the planning process is perceived to be long and difficult, the developer will often consider it advantageous to pay for an option to reserve the land. An option involves the developer paying a nominal sum to secure the right at a future date to purchase the freehold. There is usually a 'long stop' date after which the vendor is free to sell the land to anyone if the developer has not taken up the option. The option agreement might specify that after certain conditions have been complied with the developer has to purchase the land. If the developer fails to complete the purchase by the 'long stop' date then the vendor is free to market the site elsewhere. Alternatively, the agreement may allow the developer to call upon the vendor at any time to sell the site after sufficient notice. The developer will aim to fix the value of the site at the time the option agreement is entered into, but in practice this is difficult to achieve. Conversely, at least in a rising market the vendor will usually try to ensure that the open market value is fixed at the time the developer actually purchases the land.

The developer may only be able to acquire a long leasehold interest in the land at a premium with a nominal ground rent such as a peppercorn. This happens when the landowner is only able to dispose of a leasehold interest (e.g. the Crown Estate) or wishes to retain some control over the development (e.g. local authorities). The developer may be able to take a building lease in the first instance, which will immediately vest a legal estate, although it will most likely be subject to covenants relating to the satisfactory completion of the development. Alternatively, the transaction might be arranged on the basis of a building agreement and lease that, in the first instance, merely gives to the developer a licence to enter onto the site and construct the building with a commitment by the landowner to grant a lease when the building has been satisfactorily completed.

A similar arrangement is often made by local authorities in relation to freehold transactions when the developer is able to carry out the development under a building agreement and the freehold is transferred on the satisfactory completion of the development. Under this type of transaction the local authority may become an equity partner in the scheme. The value of the scheme is assessed on completion and, therefore, the authority can share in any growth. The developer can use this type of transaction to reduce risk at the outset. The developer may only be required to pay a nominal premium to enter into a building agreement and the consideration to the local authority may be only payable if a profit is made on completion. The consideration paid to the local authority may be in the form of a profit share or it may be the land value on completion. This method of acquisition is advantageous to the developer where the development scheme is large and likely to take a number of years to complete. The risk to the developer is great and the local authority, due to their interest in the implementation of the scheme, may be willing to be flexible. A prudent developer will not enter such a building agreement on a large scheme without making it conditional upon funding.

It is vital that the building agreement is carefully negotiated as otherwise long arguments can occur on completion in relation to the calculation of profit. Typically, the development scheme will take a number of years to complete and as a result it is quite likely that the personnel and the political party in overall control at the local authority will have changed by the time of project completion. This might lead to arguments over matters such as the definition of development costs that adversely affect the authority's share of the profit. An alternative to endless negotiations about how to calculate the profit is to form a joint company with the authority. With a joint company the profit is clearly shown in the audited accounts of the company and there is no argument as to what is and what is not an acceptable development cost.

Sometimes sites are acquired on the basis of a long leasehold interest with an open market ground rent payable, instead of a premium being payable with a nominal ground rent. These leaseholds are usually for between 99 and 125 years. Financial institutions prefer a term of 125 years on the basis that it permits rebuilding halfway through the term. The ground rent can be reviewed in a number of different ways and might be geared to a percentage of the open market rent of the property, a percentage of the rents received less outgoings, or rents receivable. A developer should have regard to the preferences of financial institutions. As a rule, financial institutions prefer freeholds as opposed to long leaseholds and this will be reflected in the yield at which the institution values the completed investment. The reviews might be to cleared-site value with planning permission on the assumption that a similar term of lease will be granted at the time of review. Some financial institutions prefer the revised rent of the building to be fixed at the time the ground lease is granted. This means that the ground rent might cease to rise or may even fall towards the end of the term. Ground leases may have a user covenant that will limit the use of the site to a particular planning use class.

2.7 Government assistance

What happens when land is not made available or development is not initiated by private market forces due to the fact development is not viable? Development is often not viable due to:

- low market rents;
- lack of occupier demand in a particular area;
- prohibitively high development costs as a result of the physical condition of a particular site.

Typically these areas are within the inner cities or regions in economic decline burdened with high unemployment. Very often the infrastructure in such areas is extremely congested or non-existent, and there may be contamination caused by previous use of the land by heavy industries. Government policy on urban regeneration has been 'property led': the concept that property development attracted new companies into an area, which in turn provided new jobs and benefited the wider community. There was criticism of this approach during the 1990s due to a failure to link the various government initiatives directly with the local economy of the area being targeted for regeneration. A frequently cited example is Canary Wharf in London Docklands. In response, the government launched initiatives with a greater element of competition between the areas targeted for regeneration and greater emphasis is now placed on 'partnerships' between the public and private sector. This is designed to bring about physical renewal through property development, and greater social benefits to the local community. However, this is set against the background of limited financial resources and control of public spending.

Specifically undertaken to tackle the problem, the government have introduced initiatives based on the idea of encouraging redevelopment in the inner cities by injecting public sector money into specific areas of the UK in a number of different ways through the Department of Trade and Industry (DTI). Note that powers were transferred to the devolved administrations in Scotland and Wales in July 1999 and to the Northern Ireland Assembly in December 1999. Devolution was a result of the Scotland Act 1998, the Government of Wales Act 1998 and the Northern Ireland Act 1998. However, the three Acts are different and, consequently, the devolved administrations in Scotland, Wales and Northern Ireland are different with different powers. The new Government of Wales Act 2006 came into effect in 2007 and supersedes previous legislation. Note that devolution affects the work of DTI; both the devolution legislation and the non-statutory arrangements for working with the devolved administrations have an effect. For example:

- In some cases, the legislation has transferred powers from UK ministers to the devolved administrations such that it is no longer lawful for DTI to exercise those powers throughout the UK.
- DTI needs to scrutinise new legislation made by the devolved administrations to determine whether it will affect DTI responsibilities.
- Working arrangements are set out in non-statutory documents – a Memorandum of Understanding and concordats – which create an expectation of how DTI will work with the devolved administrations.

Department of Trade and Industry (DTI) and regional development

The DTI has a strong interest in regional economic policy and has the specific responsibility for Regional Development Agencies (RDAs) in England.

Communities and Local Government (CLG) has overall responsibility within Whitehall for regional policy. DTI's approach to regional policy is designed to build the capability of regions, putting greater emphasis on growth within all regions and strengthening the building blocks for economic success and boosting regional capacity for innovation and enterprise. The DTI also has a strong interest in skills specifically in the context of:

- strengthening regional economies;
- encouraging enterprise, business growth and investment;
- knowledge transfer and innovation

DTI seeks to strengthen regional economies through sustainable improvements in the economic performance of all English regions and close the gap in growth rates between them. Successful delivery of the Regional Economic Performance Public Service Agreement (REP PSA) depends on the actions of a wide range of government departments, the RDAs and other agents that contribute to improvements in the performance of each region. The DTI is represented in each of the nine English regions by the Government Offices, which are interdepartmental bodies headed by the Regional Co-ordination Unit (for further information visit the Government Office website http://www.berr.gov.uk.

The Industrial Development Advisory Board advises DTI ministers on applications from companies who are proposing to undertake capital investment projects in the Assisted Areas in England and have applied for financial assistance under 'Selective Finance for Investment in England' (SFIE). In July 2005 the government invited the English regions to give advice on policy development and public spending decisions in relation to regional transport, housing and economic development. The government asked for advice on regions' priorities in these areas to enhance regional input into government policy development, showing how such priorities relate to each other to form a coherent, credible and strategic vision for improving economic performance of regions, and how these priorities were aligned to resources.

Selective Finance for Investment in England is designed for businesses that are looking at the possibility of investing in an Assisted Area (see below), but need financial help to go ahead. This support helps fund new investment projects that lead to long-term improvements in productivity, skills and employment. The majority of cases are appraised by the RDAs but a few, because of their size, are appraised by DTI in London. Further information is available on the Regional Investment pages of the DTI website. Finally, DTI is working in partnership with the RDAs and the Government Offices to enable policy and delivery to be led at the appropriate level and be informed by national, regional, subregional and local needs.

Regional aid and UK Assisted Areas

On the basis of Article 87(3)a and Article 87(3)c of the Treaty, state aid granted to promote the economic development of certain disadvantaged areas within the European Union is considered compatible with the common market by the European Community (EC). This kind of state aid is referred to as regional aid. Regional aid consists of aid for investment granted to large companies or, in certain limited circumstances, operating aid, both of which are targeted on specific regions to redress regional disparities. Increased levels of investment aid granted to small and medium enterprises (SMEs) located within the disadvantaged regions over and above what is allowed in other areas is also considered regional aid. In Great Britain the main forms of state aid is through discretionary grant schemes:

- Selective Finance for Investment in England (SFIE), which helps fund new investment projects that lead to long-term improvements in productivity, skills and employment.
- Regional Selective Assistance (RSA) – administered by RSA Scotland, part of the Scottish Executive, aimed at encouraging new investment projects, strengthening existing employment and new job creation.
- RSA Cymru Wales (Regional Selective Assistance) – delivered by the Welsh Assembly Government to help support new commercially viable capital investment projects that create or safeguard permanent jobs.

England's Regional Development Agencies

Eight Regional Development Agencies (RDAs) were established under the Regional Development Agencies Act 1998. They are:

- Advantage West Midlands
- East of England Development Agency
- East Midlands Development Agency
- North West Development Agency
- One North East
- South East England Development Agency
- South West of England Development Agency
- Yorkshire Forward

The ninth, in London (the London Development Agency), was established in July 2000 following the establishment of the Greater London Authority (GLA). Responsibility for sponsorship of the RDAs moved from the former Department for the Environment, Transport and the Regions to DTI in 2001. Their role is as strategic drivers of regional economic development and they aim to co-ordinate regional economic development and regeneration, enable the regions to improve their relative competitiveness and reduce the imbalance that exists within and between regions. Each agency has five statutory purposes, which are:

1. to further economic development and regeneration;
2. to promote business efficiency, investment and competitiveness;
3. to promote employment;
4. to enhance development and application of skill relevant to employment;
5. to contribute to sustainable development.

The RDAs' agenda includes regeneration, taking forward regional competitiveness, taking the lead on inward investment and working with regional partners in ensuring the development of a skills action plan where the skills' training matches the needs of the labour market. The total budget for 2006–07 for the RDAs was £2,256 million, increasing to £2,309 million in 2007–2008 (http://www.dti.gov.uk, 2007).

European Structural Funds

The European Council of December 2005 agreed a total Structural Funds budget of €308 billion for 2007–13. The UK will receive substantial Structural Funds receipts, amounting to €9.4 billion (2004 prices) during this period. These are used to meet the EU's three objectives of cohesion and regional policy, of which the Structural Funds are an instrument: convergence (European Regional Development Fund (ERDF); European Social Fund (ESF); Cohesion Fund), regional competitiveness and employment (ERDF; ESF) and European territorial co-operation (ERDF). The Structural and Cohesion Funds are divided into three separate funds:

1. European Regional Development Fund (ERDF)
2. European Social Fund (ESF)
3. The Cohesion Fund

The ERDF will support programmes addressing regional development, economic change, enhanced competitiveness and territorial co-operation throughout the EU. Funding priorities include research, innovation, environmental protection and risk prevention, while infrastructure investment retains an important role, especially in the least-developed regions.

The ESF will focus on four key areas: increasing adaptability of workers and enterprises; enhancing access to employment and participation in the labour market; reinforcing social inclusion by combating discrimination and facilitating access to the labour market for disadvantaged people; and promoting partnership for reform in the fields of employment and inclusion.

The Cohesion Fund contributes to interventions in the field of the environment and trans-European transport networks. It applies to member states with a gross national income (GNI) of less than 90 per cent of the community average, which means it covers the new member states as well as Greece and Portugal. Spain will be eligible for the Cohesion Fund on a transitional basis.

Government agencies

There are various government agencies that exist to implement and administer urban regeneration policies on behalf of the government. To make matters a little confusing there are different agencies in England, Scotland and Wales. This is often common between different states or regions in many countries. Their roles differ, but they all take an active, initiating role in the development process by making land available for development and providing financial assistance, and they may even participate directly in development.

Urban Development Corporations

Urban Development Corporations (UDCs) are non-departmental public bodies that were established under the Local Government, Planning and Land Act 1980. They were limited-life bodies tasked to secure the regeneration of their designated areas and were all wound up by the mid-1990s. In the 2003 Sustainable Communities Plan, the government stated that it was looking to establish new mechanisms in the growth areas to drive forward development, specifically mentioning UDCs in Thurrock, East London. Since then and following consultation, UDCs have been established in Thurrock (also covering the Olympic site) and West Northamptonshire. An Urban Development Area, which is similar to a UDC, has been established in Milton Keynes (under the Milton Keynes Partnership Committee consisting of a variety of public, community and private sector representatives) under the Leasehold Reform, Housing and Urban Development Act 1993. The new UDCs were established under the 1980 Act and therefore have largely the same objectives and powers as the earlier UDCs, namely:

- bringing land and buildings into effective use;
- encouraging the development of existing and new industry and commerce;
- creating an attractive environment;
- ensuring that housing and social facilities are available to encourage people to live and work in the area.

For these purposes, a UDC can:

- acquire, hold, manage, reclaim and dispose of land and other property;
- carry out building and other operations;
- seek to ensure the provision of water, electricity, gas, sewerage and other services;
- carry on any business or undertaking for the purposes of regenerating its area;
- generally do anything necessary or expedient for this purpose.

The UDCs have been invested with development control for strategic applications in support of their objectives/purpose, leaving householder and routine applications to the local authority. The UDCs do not have powers in relation to determining the overall level of development or in relation to the location and distribution of development; these are matters for the Regional Spatial Strategy and borough and district councils. The UDCs have a term set for 7 to 10 years, reviewed after 5 years, and are funded by central government. They are run by boards, the places on which are advertised through the media, but there are also guaranteed local authority representations. The boards typically comprise eleven people plus a chair and a deputy chair.

There are some important differences between these and the earlier UDCs in that the planning powers are not as extensive and the boards were appointed by, and accountable to, the then Secretary of State for the Environment. A more collaborative approach with local, regional and national stakeholders is being promoted, with the UDC working more closely with such partners, taking advantage of their resources, powers and skills where appropriate. For example, the London Thames Gateway UDC, established in May 2004, is responsible for the strategic regeneration of the Lower Lea Valley and Barking–Havering Riverside. In all, parts of six London boroughs are benefiting – Tower Hamlets, Hackney, Barking and Dagenham, Havering, Newham and Waltham Forest. The UDC is currently working on its overarching strategy for these areas along with finalising the regeneration frameworks for the Lower Lea and London Riverside.

UDCs regenerate their designated areas within a specified timetable. They aim to reclaim and service land, refurbish buildings and provide adequate infrastructure to encourage private sector development. The UDCs offer practical and sometimes financial assistance to developers and owner–occupiers looking to acquire sites in their area. They assist companies to relocate or expand in their area, with both grants and subsidies including amenity grants, rent and interest subsidies and property capital grants. Developers are able to apply for grant assistance if their proposed scheme is in a UDC area. The UDC becomes the local planning authority for their designated area, although some delegate the administration of planning applications to the local authority. It is going to be difficult to assess the success of UDCs until some years after they have finally been wound up. They were at the forefront of the government's 'property led' policies, although they have been widely criticised for ignoring social issues. There were criticisms that infrastructure projects were provided too late in the regeneration process (e.g. The London Underground extension to the Jubilee Line), due to delays and cutbacks in the provision of the necessary government funds. However, there were some successes with developments such as Albert Dock in Liverpool and The Bristol Spine Road. In defence of such criticism, UDCs point to success in bringing about physical renewal and attracting private inward investment.

English Partnerships

English Partnerships (EP) is a national urban regeneration agency set up by the government as a new initiative on urban regeneration in November 1993 (under the provisions of the Leasehold Reform, Housing and Urban Area Act 1993), and it became operational in April 1994. English Partnerships:

- develops a portfolio of strategic projects;
- acts as the government's specialist advisor on brownfield land;
- ensures surplus public sector land is used to support wider government objectives, especially the Sustainable Communities Plan;
- helps to create communities where people can afford and want to live;
- supports the urban renaissance by improving the quality of towns and cities.

EP programmes are designed to increase the supply of affordable housing and housing for key workers in areas experiencing housing pressure; make best use of the scarce supply of land by identifying brownfield land and increasing supply for development; reduce the stock of low demand and abandoned housing in the Market Renewal Pathfinder areas and areas with low demand for housing, whilst increasing the supply of new and refurbished high-quality housing and amenities in those areas. EP works with both public and private sector partners and, importantly, is able work on large and complex projects. It aims to set benchmarks for high-quality urban design, construction and environmental sustainability and act as a catalyst for development – for example, being involved early to prepare sites for development by partners. EP also devises and encourages innovative methods of dealing with problems, engages with local communities, and is involved in masterplanning, land remediation and regeneration.

The 2004 Barker Review of Housing Supply recommended that EP should have 'a lead role in delivering development through partnering with public and private sector bodies in assembling complex sites, masterplanning, remediating land and developing supporting infrastructure'. Barker noted that the government should provide greater certainty as to the principles by which EP would, or would not, intervene, to avoid crowding out private sector activity, or stunting the development of new markets. EP works with the CLG (successor to the Office of the Deputy Prime Minister) to agree the principles on which they will deliver development. EP has a role in land assembly and acquisition to deliver the government's Sustainable Communities Plan. Much of this land is owned by the public sector but at times other land has to be acquired. Where possible, the purchase is made by agreement with the landowners; however, this is not always practical. Where appropriate or necessary, EP is able to use statutory powers to acquire land compulsorily. EP depends on a wide range of external partners and stakeholders including:

- private sector companies
- The Housing Corporation
- RDAs
- local authorities
- government offices
- The Commission for Architecture and the Built Environment (CABE)
- non-governmental organisations with complementary aims.

EPs programme is delivered through a variety of delivery agencies that include:

- joint ventures with both public and private sector partners;
- specially designated urban development areas, such as in Milton Keynes;
- local delivery vehicles in the Growth Areas;
- Urban Regeneration Companies.

EP works with local communities, holding public meetings and exhibitions, local residents' steering groups, action planning events and design workshops. The planning system demands community involvement; however, EP looks at how best to incorporate this (and other models of community engagement available in regeneration and development) to develop best practice and to contribute to wider policy debates. EP investment is focused on:

- the 20 per cent most deprived areas, as defined by the CLG and currently adopted for measures, such as the stamp duty exemptions ;
- the coalfields;
- urban regeneration company areas;
- areas of major housing growth in the south-east including the four target areas (Milton Keynes and the South Midlands; the London–Stansted–Cambridge–Peterborough Corridor; Thames Gateway; and Ashford);
- housing market renewal areas, the Northern Growth Corridor and other initiatives, and strategic brownfield sites or hardcore brownfield land in, or adjacent to, any priority areas or in areas of housing pressure or abandonment.

EP can support and masterplan major regeneration exercises where brokerage expertise and powers, including the use of CPOs, can be used to pursue development and regeneration in partnership with others. EP can provide a demonstration effect through demonstration projects, illustrating the returns that might be made. Where EP owns land and assets nearby, it may acquire additional land to optimise asset value for the public sector. EP might also become involved in joint ventures with private partners, where the risks and returns from development are shared. At times, EP may bring partners together to agree on infrastructure provision and development timetables. For example, in masterplanning and enabling the development of a site where EP/the public sector takes on the development risks and returns of bringing a site to the market, the private developers take the main responsibility and risks for subsequent development. In site purchase and disposal, EP might acquire sites that are necessary to allow development to proceed. These might then be passed to the private sector at an early stage so that the developer helps to determine the shape of development. This could include the use of EPs' CPO powers.

Finally, there is gap funding where the private sector or other public partners identify projects on their land and EP funds the gap between costs and reasonable returns on the basis of public policy objectives. As such gap funding can be used to:

- fund the unlocking of key brownfield sites;
- provide infrastructure of strategic importance to a region or subregional strategy;
- provide funding to winners of development competitions;
- improve the affordability or timing of housing on existing consented sites.

Commission of New Towns

The Commission of New Towns (CNT) was a self-financed government elected agency responsible for the disposal of surplus land left after the winding up of the New Towns created under the New Towns Act 1946. The total number of New Towns created was twenty-nine spread across England. The 1946 New Towns Act envisaged that as the towns grew the development corporation would transfer any remaining assets to local authorities. In the end the CNT was formed, and from 1961 it was responsible for managing and disposing of the land and property assets of the defunct English development corporations. In May 1999 CNT merged with the Urban Regeneration Agency to create English Partnerships, the government's current national regeneration agency.

The early 1960s saw rising births and predictions of an increase in population of some eighteen million by 2000; the New Towns Act was re-invoked and nine New Towns were designated during the 1960s, including the growth of Peterborough, Northampton and Warrington, and the designation of the 'new city' of Milton Keynes, the largest 'greenfield' project. The programme culminated in 1970 with the designation of Central Lancashire New Town.

The English New Town programme represented one of the most successful urban policies of post-war Britain and has housed over two million people, provided over a million jobs, and has evolved economically and socially successful communities. Milton Keynes is, for example, considered to be one of the most successful large-scale development projects ever undertaken in England, delivering homes, jobs, facilities and services to create a balanced community. New settlements are one of the options considered for managing growth and towns like Milton Keynes provide valuable templates for future large-scale development. The lessons provided by the New Towns are being reassessed and used to help deliver the Sustainable Communities Plan. Regeneration is being brought about through a range of initiatives led by EP in collaboration with local authorities and other stakeholders, to bring the New Towns up to date. Initiatives such as the Millennium Communities Programme, Urban Regeneration Companies, the security permission required Town Centre Renewal Programme and sustainable urban extensions (e.g. Upton, Northampton) are all revitalising New Towns.

Welsh Department of Enterprise, Innovation and Networks

The Welsh Development Agency (WDA) was an executive agency charged with encouraging business development and investment in Wales. In April 2006, the WDA together with the Wales Tourist Board, and Education and Learning Wales (ELWa) merged with the Welsh Assembly Government. WDA functions are now carried out by the Assembly's Department of Enterprise, Innovation and Networks (DEIN). The WDA worked to encourage entre-preneurial growth in terms of the annual number of start-up businesses based in Wales and to persuade multinational companies to relocate or open subsidiary facilities in Wales. This responsibility has been devolved to DEIN's International Business Wales organisation. The service supports a range of options, from identifying premises for immediate occupation to facilitating refurbishment, extension or new build. The service covers information, advice, practical assistance, project management and the provision of customised properties. This includes:

- Aasearchable database of all available properties;
- business parks to meet the needs of specific sectors and high-growth firms of all types;
- property experts to find premises or sites for development;
- land development experts specialising in the acquisition and redevelop-ment of large brownfield and strategic sites for future development;
- a full project management service from conception to supervision of construction.

A Property Development Grant is available to private sector companies and financial institutions undertaking new development or upgrading/extending existing business sites and premises for either speculative or bespoke schemes where there is a deficit between the costs of the projects and the value of the completed development. Eligibility requirements are:

- use – any use of land or premises for business purposes including industrial, warehouse, office, retail and the commercial element of mixed-use development;
- location – small and medium-sized enterprises anywhere in Wales, large enterprises Assisted Areas only;
- interest in land – freehold or a lease of 10 years or greater;
- eligible land expenditure – the lower of actual land purchase price or open market value;
- the costs of providing services, utilities and infrastructure;
- construction costs – evidence of competitive tendering may be required;
- site preparation;
- professional fees associated with the project;
- finance charges – interest or similar financial charges will be allowed subject to independent verification that they are in line with generally available market terms;
- developer's profit – developments by the owner–occupier will not be eligible for development profit.

There is a two-stage approval process, comprising 'Approval in Principle' (this provides an indication of profit eligibility, viability, preliminary level of grant and availability of budget) followed by 'Final Approval', which defines the grant amount and the terms and conditions. Recipients are required to enter into a formal, binding legal agreement with the Assembly. Conditional upon the information being provided, the Assembly will provide 'Approval in Principle' within one month of receipt of the application for projects with a total expenditure below £500,000 and two months for larger projects.

Scottish Enterprise and Local Enterprise Companies

Scottish Enterprise (SE) is the main economic development agency for Scotland, covering approximately 93 per cent of the population. Scottish Enterprise consists of Scottish Enterprise and twelve Local Enterprise Companies. Working with the private and public sectors, SE seeks to secure the future of the Scottish economy by making the industries more competitive by helping business start-ups and existing companies to grow; promoting and encouraging exporting; attracting inward investment and developing skills. SE is funded by the Scottish Executive with headquarters in Glasgow, and was established by an Act of Parliament on 1 April 1991 when the Enterprise and New Towns (Scotland) Act 1990 came into force. The Act merged the Scottish Development Agency and the Training Agency in Scotland. SE is a statutory body, known as an 'Executive Non-Departmental Public Body', which means SE is a public body but not a government department or part of one. SE has its own board appointed by Scottish ministers and a chief executive who is the Accountable Officer responsible to the Scottish Executive.

Funding and grants

Direct government financial assistance via funding or grants is available to local authorities and/or developers proposing urban regeneration schemes in specific areas of the country. There are a number of financial grants available to developers and the public sector to develop derelict or rundown inner city sites and buildings from the government and EU. We shall now examine the various types of funding and grants available.

European grants

At present, of the four Structural Funds allowing the EU to grant financial assistance to resolve structural economic and social problems, the European Regional Development Fund (ERDF) is the most relevant. The ERDF's principal objective is to promote economic and social cohesion within the EU through the reduction of imbalances between regions or social groups, as is the European Social Fund (ESF), which is the main financial instrument allowing the EU to realise the objectives of its employment policy.

The ERDF is the fund for major industrial development and infrastructure projects. All EU structural funds are administered in accordance with five priority objectives known as Objectives 1 to 5. ERDF is available under Objectives 1, 2 and 5(a). In practice, all development areas are covered: transport, communication technologies, energy, the environment, research and innovation, social infrastructure, training, urban redevelopment and the conversion of industrial sites, rural development, the fishing industry, tourism and culture. ERDFs were available from 1993 to 1999 and 2000 to 2006. Regional policy support between 2000 and 2006 were:

- Objective 1 – development of the least favoured regions;
- Objective 2 – conversion of regions facing difficulties;
- Interreg III – interregional co-operation;
- Urban II – sustainable development of urban areas;
- innovative actions – development of innovative strategies to make regions more competitive;
- transport and the environment in the applicant countries.

For example, Merseyside, and the Highland and Islands in Scotland received ERDF funds, as Objective 1 locations until 1999. Areas with Objective 2 status included many of the industrial regions in the North, Wales and Scotland and some London Boroughs. Objective 5(a) areas were underdeveloped rural areas. Belfast received funding under the Urban II initiative.

ERDF money spent has to be matched pound for pound from other sources, i.e. the fund will only supply up to 50 per cent of the cost of a project, and cannot be used to replace government funding.

Enterprise Zones

From 1981 the government designated areas of the UK as Enterprise Zones to encourage private sector industrial and commercial development through the removal of administrative controls and the provision of tax incentives. By 1990, thirty-two Enterprise Zones were designated, the first being the Lower Swansea Valley in 1981. Enterprise Zone status lasted for 10 years and all zones have now lost their designation, the last being Tyne and Wear in October 2006.

Simplified Planning Zones

Simplified Planning Zones (SPZs) are designed to speed up the rate of development in areas where it is needed. Local authorities can use SPZs to give advance planning permission for specified types of development in clearly defined areas. The advantage to developers is that they know in advance what schemes will be permitted and that they do not need to make a planning application or pay planning fees. The first local authority to introduce an SPZ was Derby. Like Enterprise Zones, their status lasts for only 10 years. Any planning authority can propose an SPZ and the only requirement before formal adoption is to publish a draft scheme for public comment. If objections are raised there may be a public inquiry. A developer can propose an SPZ, and if the authority rejects the proposal the matter can be referred to the Secretary of State for the Environment.

Land availability

The UK government is concerned about the amount of vacant and derelict land within urban areas, and actively seeks to promote brownfield or development on previously used land.

Land registers

In 1980 under the Local Government, Planning and Land Act the government introduced land registers. Local authorities are required to provide the Department of the Environment with a register containing details of unused and underused land owned by public authorities. The register is available for inspection for potential purchasers and developers at the Department of the Environment's regional offices. It contains a description of each site with information on ownership, planning history and constraints to development. The Department of the Environment can assist developers to obtain the release of particular plots of land on the open market. However, the presumption of the government in setting up the register was that ownership was the greatest constraint. In practice developers will approach landowners if a viable development is possible. The greatest constraint preventing many areas of vacant land being developed is lack of occupier demand and the physical condition of the site.

On 13 October 2003 the Land Registration Act 2002 and Land Registration Rules 2003 came into force. They have completely replaced the law for land registration. Together, the new Act and Rules govern the role and practice of the Land Registry. The Land Registry compiles a register title to land in England and Wales and records dealings (for example, sales and mortgages) with registered land.

Around eighteen million titles were entered onto a computer database in 2002 and are now fully computerised. The Land Registration Act 2002 achieves the following:

- it simplifies and modernises land registration law and is the first major overhaul of the land registration system for 75 years;
- it makes the register a more complete picture of a title to land – showing more fully the rights and interests affecting it;
- it provides a framework for the development of electronic conveyancing.

The major changes to the law and practice affecting those involved with registered land were:

- shorter leases must now be registered;
- voluntary registration is available for new types of interest in land;
- changes affect the protection of third-party interests;
- the law of adverse possession (squatters rights) has been reformed.

Public Request to Order Disposal

In 1993 the government launched an initiative to bring public vacant land into use, known as 'Public Request to Order Disposal' (PROD). The aim is to unlock the development potential of 80,000 acres (32,374 ha) of publicly owned vacant land. PRODs are a useful but little used legal power that anybody can use to force the sale of publicly owned empty homes or abandoned land in England and Wales. The power dates from the 1980 Local Government Planning and Land Act. It enables anybody to request that the Secretary of State investigate why publicly owned properties or land have been left empty. The Secretary of State can order public authorities to dispose of empty homes or land. Where this happens, the land or homes are usually sold to the open market.

This legal power covers most but not all publicly owned property in England and Wales but does not operate in Scotland or Northern Ireland; however, the government will investigate any vacant public property brought to its attention that is covered by the legislation and may investigate cases where the legislation does not apply. As of 1 April 2005 there were over 104,000 empty properties in the public sector, including housing association stock. This sum represents 15 per cent of the total number of empty homes in England (approximately 680,000). Note the government does not report the number of long-term vacant properties (i.e. properties empty for more than six months) owned by the public sector; estimates put this figure at around 50,000.

Discussion point

In what ways do governments assist the development process?

Reflective summary

This chapter has shown that even if a developer has a well-researched land acquisition strategy, its achievement, in the way and time originally envisaged, is very often beyond the control of the developer.

The following preconditions need to be in place for the development process to be initiated through land acquisition:

• the landowner's willingness to sell the land on terms and at a price to enable a viable development to proceed;

- planning permission for the proposed development or allocation of the proposed use within the relevant development plan;
- the existence of infrastructure and services to support the proposed development;
- the existence (if necessary after appropriate treatment at a reasonable cost) of suitable ground conditions to support the development;
- the necessary development finance;
- a known end-user or occupier demand for the proposed development.

If any one of the above preconditions is not in place then development will not proceed or subsequent development will represent a considerable risk to the developer. Local authority involvement and government assistance may be available in relation to compulsory acquisition, provision of infrastructure, site reclamation, finance or occupier/investor incentives depending on the nature of the proposed development and its location. Most importantly, the requirements of occupiers must be at the forefront of any developer's land acquisition strategy.

Case study: Brunel Quays, Lostwithiel, Cornwall

This case study of a residential development in Lostwithiel in Cornwall, undertaken by Wombwell Homes, illustrates many of the aspects of the initiation and acquisition stages in the development process examined in this chapter. The site was in a derelict state and used formerly by the Great Western Railway Company for storage and sidings. The site also had some residual contamination from its former use when it was bought by the developer.

The developer

Wombwell Homes is an independent residential developer that specialises in brownfield developments, and renovation and redevelopment of existing buildings. The company started in 2000 and is relatively small, comprising ten people who include an accountant, a personal assistant, a contracts manager/administrator, a secretary, a sales manager and sales assistant, a construction director, a quantity surveyor, a management company administrator and a handyman.

Site acquisition

The site is located in Lostwithiel, once one of the largest tin exporting ports in Cornwall and the Middle Ages' capital of Cornwall. The site occupies a central location, on the banks of the River Fowey. The site was beset with development problems, due to its location, sharing boundaries with the river on one side, and the Paddington to Penzance Railway line on the other. Access to the site is restricted through the narrow station approach, between the Scheduled Ancient Monument of Lostwithiel Bridge and a busy level crossing, whose barriers are down over twenty-five times a day.

The buildings on the site were designed by engineer Isambard Kingdom Brunel in the nineteenth century as part of the Great Western Railway. The railway line has ceased to operate and the buildings decayed over time. Brunel Quays is a classic example of the regeneration of a historic and Grade II listed brownfield site, being part renovation of the Brunel buildings and part new-build construction of a sympathetic redevelopment and renovation.

Over time the site had become a tipping ground for rubbish. There was a derelict house on site used as a squat, which had become a magnet for the local drug culture. Other buildings were in an advanced state of dereliction and unfortunately had become a local eyesore in the centre of the town.

The site was zoned for residential use. It was known there was some 'small level and localised' contamination resulting from the former use and this included the existence of mercury and arsenic. A concern was the proximity of the river and the potential for migration of the contaminants. At the time of acquisition the value of the site was £150,000 without planning consent and with planning consent £750,000. The overall site size was approximately two acres with 7,500 ft^2 of building plus a burnt-out section of building comprising approximately 3000 ft^2.

Initiation

The developer specialises in residential development and did not consider any alternative use for the land. The demand for housing in the UK was high during the late 1990s and in the early 2000s, especially in the South West where the acquisition of second homes for rental as holiday homes and as retirement homes was becoming popular. The land was advertised for sale with zoning for residential use.

The Brunel Quays project had been on the drawing board for nearly 10 years, and when the site received an outline planning approval, albeit for an unsuitable scheme in the developer's view, the principle of a residential housing scheme being allowed on a listed industrial site had been established. The time to progress the project was right and Wombwell Homes, who had established a track record in redevelopment of derelict and brownfield sites in the local area, approached the local planning officer.

The ground investigation confirmed the presence of some contaminants. Given that the land was for residential use it had to be cleaned to a very high standard. The

Plate 2.1   Brunel Quays derelict, abandoned, original buildings on the site. Reproduced with permission of Wombwell Homes.

developer had to consider the impact of the contamination and whether the siting of the new buildings might cause them to contain some of the contaminants on the site.

Implementation

From the outset, every aspect of this difficult site threw up new problems or challenges. The site comprised made-up ground covering the old river beds and harbour that had made Lostwithiel one of the top five biggest ports in England and the medieval capital of Cornwall. With the decline of tin, the entire area was filled in for the construction of the railway yard, sidings and repair sheds.

The site was raised over one metre higher than the town on the other side of the river, and had never flooded. The location alongside the river brought up the flood plain issue, and although the Environment Agency issued a flood plain layout of high-risk areas that excluded the site initially; before a planning approval application was submitted, the site became included, which was interesting given that the site had never flooded even when the majority of Lostwithiel had. The decision added many thousands of pounds to the cost of construction, as all levels had to be raised by 500 mm, the expensive piling and foundations had to be extended, and retaining walls had to be introduced, which needed to allow for the now increased finished floor levels to be achieved and made workable.

When work commenced on site, the small and insignificant pockets of contamination, highlighted by the desk study and early trial pits, transformed into much higher concentrations of contamination. This led to a consultant-designed containment scheme over most of the site, incorporating expensive gas barriers and specialist membranes being installed. Anti-leaching trenches, 30 m × 3 m × 3 m, were dug and filled with imported material to stop contamination entering the river. Removal of excess excavated material had to be kept to a minimum, as the cost was over six times the norm, and the material had to be taken to a specially licensed tip, which was not located as close as the usual one. When all the fees, additional work and time taken in remediation of the site were added up, the cost exceeded a quarter of a million pounds.

Drainage was a problem, the only way offsite was uphill, and the nearest sewers were hundreds of metres away and the wrong side of the river, the road or the railway line. After months of meetings trying to get the most efficient, cost-effective drainage scheme, the developer was granted approval to discharge into an existing sewer that went under the railway line and was connected to the main sewer beyond. Apart from reforming the manhole to receive the new drain discharge pipe, various surveys were carried out, and a closed-circuit television (CCTV) scan of the existing pipe resulted in a specialist 'lining' repair being completed. However, the main problem with this point of discharge was that it was higher than the houses that were being built. In addition, the only way the developer would be allowed to discharge into the existing drain, due to its restricted diameter, was under a controlled flow rate. Taking these things together, all the developer could do was to pipe the drainage by a gravity system from the top of the site, down to the lowest point, where an overflow/storage chamber and pumping system was installed.

Water and telecom services installation were simple, both having mains at the site entrance. Gas and electric were problematic and expensive at approximately £100,000. No gas infrastructure had been on site, and the nearest mains were the wrong side of the river or railway line. Consultants debated whether the gas main could be strapped below the railway bridge, drilled under the river or 'moled' under the railway (the feasibility study itself cost £6,000).

There was electricity on site and the pole-mounted, high-voltage electricity main was visible only twenty metres away, on the other side of the railway line, and the developer had an existing underground connection in place from it on the site. It was hoped to use this main for the site, in an upgraded form if necessary. The electricity company had no record of this sub-main, and although adequate for the project and in full working order, they refused to accept it – a new one laid by them would have to be installed. This brought Network Rail into the equation (Railtrack at the time), as the developer had to gain access to a mainline, trackside substation on Network Rail land, to make a new connection into an electrical system that powered the signal box and all railway electrics around the level crossing and station area. Firstly, they had to establish that the sub-main was ducted under the lines to the site before even contemplating complying with the onerous regulations stipulated by Network Rail (which included a fine of £1,000,000 per day for any interruption to main line services!). Months of correspondence and phone calls produced no results, and the developer finally lost patience and used the electric company's influence to get into their substation. It was obvious that the main under the lines was not ducted, and the state of the electrics within the substation was so poor that a new problem became evident. It was deemed prudent to leave well alone; with no duct available, the main unusable and in a poor state of repair, a new route was required, and the nearest place to make the connection, although over 600 metres away, and well offsite, was at least clear, with no river or railway line between.

Whilst the gas board pontificated, and the electric company became embroiled with land owners for compensation deals and way leaves, the construction on the site had progressed. In the developers words, the first block of twelve apartments had been:

> built on foundations that would withstand a nuclear holocaust; had every gas barrier known to man installed within them; they had received doors and windows shipped from Sweden, granite shipped from Portugal and slates shipped from China, but still no sign of any gas or electric appeared on site, from much nearer to home, and now occupations were imminent!

Time was critical. The electrical substation was installed, but the cable needed to make it live was 600 metres away and with work in between having started, things got tense. The painters were finishing, carpets were being laid, owner's removal dates were fast approaching, and the developer seemed powerless to avoid a disaster with disappointed purchasers. They could move into their new homes, but they couldn't turn on a light, boil a kettle, or switch on the heating. Threats of ombudsman involvement prompted contact from a senior executive who agreed their service had been abysmal and promised to resolve the situation. This was doubtful, as Network Rail had a mandatory period of notification required before anyone could drill any service under the line, and it appeared the developer had run out of time. Someone was doing something but was it too late? Meanwhile, electricity trenching had begun, and four days later the 600 metres had been breached in a feat by working around the clock – Western Power Distribution delivered, and with less than a week before occupation the switch was flicked and the lights came on.

What about the gas? After six months of silence and inactivity, the senior official acted, bypassing Network Rail, their timescales and regulations in a swift proclamation. Due to an anomaly in the law, the Highway Agency had authority where a railway line crosses a road at a level crossing (and it is always this way, never does a road cross a railway line). This having been established, the gas board have complete authority to dig up, or tunnel under the roads, without notice, for emergency gas works, which they deemed this to be.

Three days before the first occupation, no gas connection had been made and the situation was no different from what it had been at the outset. As with the electricity, all on site pipework, ducting, cabling and trenching had long since been completed, but

Plate 2.2 Brunel Quays completed development. Reproduced with permission of Wombwell Homes.

Plate 2.3 Brunel Quays from the River Fowey. Reproduced with permission of Wombwell Homes.

Plate 2.4 Brunel Quays completed apartments. Reproduced with permission of Wombwell Homes.

Plate 2.5 Brunel Quays completed apartments. Reproduced with permission of Wombwell Homes.

a 50-metre gap still existed with a railway line and level crossing in between the end of the on site gas pipe and the end of the live gas main. Few people saw the orange waist-coated men, bearing fencing, coloured barriers, bollards and night-working lights, as they set up, on that out-of-hours Friday night. Few people were aware as the end of the gas main was exposed, and a remotely controlled 'mole' tunnelled beneath the level crossing and into the site, allowing the new gas main to be fed through. By the Monday morning, the gas was on and the preceding six months had vanished, as if they'd never happened.

Work commenced on site in February 2003 and finished May 2005. The final development comprised a three-phase project:

- Phase 1 – twelve apartments and one house, and the conversion of the burnt-out building into eight units;
- Phase 2 – offices, and fourteen apartments and houses;
- Phase 3 – twenty apartments.

Ninety-five per cent was sold off plan and the remainder sold at completion. The apartments started at selling at £119,000 and in 2007 were selling for £220,000. The houses started at selling at £189,000 and in 2007 were selling for £250,000.

The final scheme is slightly different to that originally approved by the planning authority. The café was removed from the scheme and the Section 106 was amended. The planning process was considered long and tortuous by the developer; however, the reason was that this development was unique in Lostwithiel and the planners had not dealt with any similar applications. In addition the site is very prominent within the town.

Conclusion

After the completion of the fifty-fourth and final property, one sees the established, landscaped gardens and lawns of Brunel Quays with its natural stone, granite and balconied houses reflected in the glistening Fowey River alongside. Little is remembered of the early problems of flooding, contamination, drainage, rivers, railways and services. Three plaques proclaim the prestige and high regard this triple award-winning project has attained. Wombwell Homes are proud of the accolades as they have produced a high-quality development that can only encourage further investment in the area. However, there is no new bridge over the river, or working café; the echo of an enormous and unprecedented council U-turn and other risks associated with development. Wombwell Homes believe the development provides a good example of urban regeneration, showing what can be achieved if all the stakeholders in the property development work together, problem solve, and are prepared to be flexible.

CHAPTER

Development appraisal and risk

3.1 Introduction

In this chapter we shall examine the way in which development projects are evaluated, focusing upon the process of development appraisal and valuation. Assessing and evaluating a development is a constant process where the developer must not just conduct a one-off appraisal prior to the acquisition of a development site but rather needs to constantly reappraise the profitability of the scheme throughout the development process due to the effect of myriad influencing factors.

Risk is an inherent part of the property development process and we shall consider how this is assessed as part of the evaluation process. Market research is an integral part of the appraisal process and we consider this further in Chapter 7. In this chapter, we discuss the conventional technique of development appraisal before introducing the various cash flow concepts including the discounted cash flow (DCF) approach. Finally, we shall examine the influence of uncertainly and how this can be contained in order to reduce risk.

3.2 Financial evaluation

The conventional technique

Conventional techniques of identifying the various components of value in a proposed development are relatively straightforward using a form of 'residual' valuation. This type of model is designed to isolate the components of a development, such as risk/return or land value, and assess their individual 'unknown' value when information about the other variables are known.

Quite simply, it starts with the 'known' or certain variables and then moves to the 'unknown' variables to complete the analysis. It commences with the total value of the completed project and then deducts selected components to arrive at the remaining or 'residual' component. The main variables in the model are:

- land price
- building construction cost
- prevailing rents/prices
- interest rates
- investment yields
- time

There are two primary types of residual valuation depending on the final outcome sought.

1. **Investment risk/return**   Total development costs (e.g. land, construction cost) can be deducted from the estimated value of the completed development to establish whether the project produces an adequate rate of return for the developer or financier, either in terms of a trading profit, an investment yield or return on capital.
2. **Affordable land price**   A residual valuation is also used to assess the likely costs of producing a development scheme and by deducting these costs from an estimate of the value of the completed development scheme to arrive at an affordable land price. The developer will include an allowance for the required return in assessing the total development costs.

The manner in which the above mentioned variables are brought together in a development appraisal can be shown best via a working example.

Let us assume a 0.8 ha site (2 acre) is on the market and the vendors are seeking a price of £3,500,000. The site is in a good location in a town with adequate transport access and the vendor has obtained planning consent for 3,553 m² (38,250 ft²) of offices. Through market research, a developer has established that rents are currently £290.62 per m² (£27 per ft²) for comparable office space. A bank has agreed to provide short-term finance for the scheme at an interest rate of 2 per cent above the bank's base rate of 6.25 per cent, to be compounded quarterly, i.e. an effective annual rate of 8.47 per cent. The developer's quantity surveyor has advised him that building costs are currently £1,291.66 per m² (£120 per ft²). The agents have advised the developer that the completed scheme should achieve a yield of 7.0 per cent when sold to an investor. The developer will carry out the following typical conventional evaluation as shown in Example 3.1. Note there may be rounding errors in some of the calculations when totalled.

We shall now examine each of the elements of the appraisal in detail.

Example 3.1   Evaluation of profit (risk/return)

	£	£
(a) Net Development Value		
(i) Estimated Rental Value (ERV)		
Net lettable area 3,553 m² (38,250 ft2) @ £290.62		
per m² (£27per ft²)	1,032,573	
(ii) Capitalised @ 7.0% YP in perpetuity	14.29	
	14,751,041	
(iii) less purchaser's costs @ 2.75%	405,654	
Net Development Value (NDV)		14,345,387
(b) Development costs		
(c) Land costs		
Land price	3,500,000	
Stamp duty @1%	35,000	
Agent's acquisition fees @ 1%	35,000	
Legal fees on acquistion @ 0.5%	17,500	
		3.587,500
(d) Building costs		
Estimated building cost		
Gross area 45,000 sq ft (4,181 sq m) @ £120 p.s.f.		
(£1,291.66 p.s.m.)	5,400,000	5,400,000
(e) Professional fees		
Architect@ 5%	270,000	
Structural engineer© 2%	108,000	
Quantity surveyor @ 2%	108,000	
M & E engineer @ 1.5%	81,000	
Project manager @ 2%	108,000	
		675,000
(f) Other costs		
Site investigations - say	17,500	
Planning fees - say	6,000	
Building regulations	30,000	
		53,500
(g) Funding fees		
Bank's legal/professional fees - say	60,000	
Bank's arrangement fee	90,000	
Developer's legal fees - say	50,000	
		200,000

Example 3.1   continued

(h) Finance costs

(i) Interest on land costs (£3,587,500) over the
development period and void of 30 months @
8.25% compounded quarterly = (1.0206)   812,513

(ii) Interest on (d) building costs, (e) professional
fees, (f) other costs and (g) funding fees divided by
a half (£6.33m/2) over building period of 12 months
@ 8.25% compounded quarterly = (1.0206)4   269,238

(iii) Interest on building costs, professional fees,
other costs and funding fees (£6.33m) over void
period of 12 months @ 8.25% compounded
quarterly = (1.0206)4   538,477
                                                       _____
                                                                        1,620,228

(i) Letting and sale costs

Letting agents @ 15% ERV   154,886

Promotion   95,000

Developer's sale fees @ 1.5% NDV   215,181

Other costs (see text)
                                                       _____
                                                                          465,067

(j) Total Development Costs (TDC)   12,001,295

(k) Developer's profit

Net Development Value   14,345,387

Less total development costs   12,001,295
                                                 _____
Developer's profit   2,344,092

Developer's profit as % of total development costs   19.53%

Yield on development cost   8.60%

Net development value

There are two important variables in establishing a development scheme's value, namely 'rent' and 'investment yield'. Both variables can have an adverse effect on the appraisal model and it is critical that careful attention is paid to ensure that both are accurate.

1. **Rent**   The anticipated rent that a tenant is likely to pay to occupy the proposed scheme is usually established in consultation with an agent or a valuer, and should reflect the interaction between supply and demand

factors in the property market at a certain point in time. An estimate of the rent must be as realistic as possible and based on a thorough analysis of the present and future market trends – it is often referred to as 'fair market rent'. According to the International Standards Valuation Committee (2005), market rent can be defined as:

> the estimated amount for which a property, or space within a property, should lease on the date of valuation between a willing lessor and a willing lessee on appropriate lease terms in an arm's-length transaction, after proper marketing, wherein the parties had each acted knowledgeably, prudently, and without compulsion.

The best source of information about the current market levels of rent is to refer to comparable evidence based on recent lettings of similar schemes within the town or surrounding area – as per the definition of fair market rent stated above. It is important that each lease complies with each section of the definition and in all respects is representative of a normal rental agreement. In other words, the tenant involved in the comparable lease must have been fully conversant with the market and may have even leased the proposed development in question (if it was completed). As no two properties are identical, it is important that the property involved in each rental lease is adjusted to reflect differences in age, quality and specification. For instance where there are few (if any) recent comparative leases, a thorough market research exercise needs to be undertaken (see Chapter 7). It is critical that the developer bases the estimate on firm reliable evidence and careful analysis to establish today's rent level, i.e. at the date of the appraisal. The importance of this variable can not be overemphasised, as an overestimation in rental levels can result in a false presentation of potential risk and return. Note that it is not recommended that a developer should rely heavily on a forecast or overestimated rent in this type of appraisal and it can be argued that a forecast rent should be explicitly stated in a valuation model (DCF or capitalisation of income approach) rather than bound up in the discount rate or all-risk yield. The property market is a complex interaction of variables and it is extremely difficult, if not impossible, to forecast future rent levels with a high degree of accuracy – without doubt the emphasis should be placed on the existing market and the current level of agreed rents.

Rents are usually analysed by reference to a rate per square foot (or per square metre), based on the cost per year. In the case of an office building, the net area of the building (the internal usable space excluding circulation space and toilets etc.) needs to be established and is known as the net lettable area (NLA). The tenant is most interested in the NLA available, and tenants seeking a larger area to lease will often be rewarded with a smaller rate per square metre rate. Larger rental areas are usually cheaper (on a 'per unit of area' basis) than smaller rental areas due to bulk discount, even though other influencing factors (e.g. use, building, floor level) are identical. In addition, discounts will be applied to the rate used to take account of areas with less demand, such as basements (due to no or restricted natural light) and floors with restricted headroom. At times it can be difficult to ascertain what components on a floor should be included in the NLA and what is excluded, and this can have an adverse effect on the aggregate amount of NLA (especially in a multi-level office or retail building). All measurements should be made in accordance with the guidance contained in the current Royal Institution of Chartered Surveyors Code of Measuring Practice, which sets out the various definitions of measurement.

With an industrial scheme it is usual for the 'gross internal area' to be measured, which includes all the internal space within the external walls. Retail units are an exception, where the rent is usually analysed in relation to the traditional 20 ft (6.1 m) zones measured from the front of the shop. Note the width of the zones varies between regions – for example, some parts of London use 30 ft (9 m) zones and some parts of Ireland use 15 ft (4.5m) zones. The first zone is known as Zone A, the second Zone B and the remainder Zone C. Retail units are valued in relation to the Zone A rent, which is the rate applied per square metre (or per square foot) to the area of Zone A. Then half the Zone A rate is applied to the area of Zone B and a quarter of the Zone A rate is applied to the area of Zone C. This zoning is to reflect the fact the most valuable space is at the front of the shop. As most shops do not conform to a standard size and shape, adjustments will be made to the above described analysis to reflect the varying levels of demand for different frontages and unusual shapes, which largely will be based on the experience of the valuer and their knowledge of the prevailing market. Large retail units such as department stores and variety stores, together with shops in small parades, are usually analysed on an overall rate rather than any Zone A rate.

2. **Investment yield** The return on the total capital outlaid to purchase a real estate investment is usually referred to as the 'investment yield'. Used in the process of assessing value, the investment yield is used to discount the future rental income stream in order to calculate the capital value of the development scheme in today's money. From the investment yield a multiplier is derived that can be applied to the future income stream and this is known as the year's purchase (YP) in perpetuity. The YP is the reciprocal of the investment yield.

In effect, this approach is based on taking a snapshot of the current rental income and, where this represents current market value, assumes that it will remain at its present level in perpetuity. The growth in the rental income and the risks associated with it are then reflected in the multiplier used. For example, if the future income from the scheme is estimated to be £100,000 per annum and the investment yield is 5 per cent then 5 is divided into 100 to calculate a YP of 20, which is then multiplied by the rental income to produce the capital value of the development – in this case £2,000,000. Therefore the investment yield (5 per cent) is derived by dividing the rental income into the capital value (0.05) and then multiplying by 100 to express it as a percentage. Table 3.1 below lists a range of investment yields and their respective YP.

Table 3.1   Investment yields and respective YP

Investment return (%)	Years purchase (YP)
30.0	3.3
27.5	3.6
25.0	4.0
22.5	4.4
20.0	5.0
17.5	5.7
15.0	6.7
14.0	7.1
13.0	7.7
12.0	8.3
11.0	9.1
10.0	10.0
9.0	11.1
8.0	12.5
7.0	14.3
6.0	16.7
5.0	20.0
4.0	25.0
3.0	33.3
2.0	50.0
1.0	100.0

The investment yield is obtained by undertaking, by way of comparison, an analysis of recent sales of properties that are similar to the development scheme being proposed. When undertaking the comparison, consideration should be given to a cross section of variables that affect the value of each property – a starting point for this comparison is to consider the two main variables, the land component and the building component. The yield itself is a measure of the property investor's perception of the future rental growth against the risk of future uncertainty. In general terms the faster the rent is expected to grow, the lower the yield an investor is prepared to pay at the outset. However, the level of perceived risk in an investment can have an adverse effect on the property itself and, therefore, risk factors also need to be taken into account. The level of yields tends to change with changes in the patterns of rental growth or investor demand. Generally, shop developments tend to attract the lowest yields while industrial developments tend to attract the highest yields, due to their previous history of rental growth.

3. **Purchaser's costs**   Regardless of whether the developer intends to hold the property as an investment or sell the property on completion, the development value needs to be expressed as a net development value to allow for purchaser's costs such as stamp duty, agent's fees and legal fees (including VAT).

Development costs

1. **Land costs**   Land costs include the land price, which is either the price already negotiated with the landowner or, as in this example, the price being sought by the landowner.

   Site acquisition costs and fees normally comprise stamp duty at a percentage of the land price, legal fees involved in acquiring the site and any agent's introduction fee. In 2006 stamp duty rates for non-residential land was divided into four bands: for land priced up to £150,000 no stamp duty is payable; for land priced between £150,000 and £250,000 a payment of 1 per cent is levied; for land priced between £250,000 and £500,000 a duty of 3 per cent is charged; and for land exceeding £500,000 a 4 per cent stamp duty is payable. Note that for the purposes of the example the stamp duty rate is assumed to be a flat 1 per cent. Legal fees are usually between 0.25 and 0.5 per cent of the land price, depending on the complexity of the deal. Agent's fees are normally agreed at 1–2 per cent of the land price, depending on whether the agent is to be retained as the letting and/or funding agent.

Complications may arise in certain jurisdictions with regards to stamp duty, property tax and land tax (if applicable). For example, if the project is being forward-funded by a financial institution, the developer may have to allow for double stamp duty on the land cost if the developer purchases the site before completing funding arrangements with the financial institution. Stamp duty will be incurred on the initial purchase of the land by the developer and the subsequent transfer to the financial institution.

2. **Building costs**   Building costs are estimated by the developer's quantity surveyor and are usually expressed as a 'per unit' cost, such as the overall rate per square metre (or square foot) – this 'per unit' cost is then multiplied by the gross area of the proposed building. The building costs are estimated at the time of the proposed implementation of the development project. Usually no allowance is made for inflation during the building contract period but some developers may inflate building costs in their appraisals, particularly in periods of rapidly rising building costs or if the construction period covers an extended period of time and there is certainty surrounding future building costs.

3. **Professional fees**   These fees are normally calculated as a percentage of the building costs and include all fees for professional services employed in the completion of the development – most often this includes the architect, the quantity surveyor, the structural engineer, the mechanical and electrical engineers, and the project manager. The actual rates per professional can vary considerably in accordance with factors such as the size of the project, the capital outlay and complexity of the task. Where the total cost of professional fees are not standardised they are normally in the vicinity of 12–13 per cent of the building cost – they are either calculated on a 'flat fee' basis or, alternatively, based on the scale charges of each profession, a negotiated percentage or a fixed fee. The percentage agreed with each member of the professional team depends on factors such as the nature and scale of the development, as well as the relationship/goodwill between the developer and each professional. Small refurbishment schemes normally attract higher percentages than larger, complex development projects. Perceived high profile or 'blue ribbon' developments may cause competition between professionals who are more than willing to be a part of the project, which in turn may equate to a lower rate. If a developer is to appoint other professionals, such as a traffic engineer, a landscape architect or a party wall surveyor, then these need to be included in any evaluation of the project.

4. **Site investigation fees**   These include fees for ground investigation and land surveys. Especially for residential developments, any land contamination must be identified and rectified – there are many examples of a building being constructed on contaminated land (not identified in the initial site survey), which necessitated the new building to be demolished prior to the site being decontaminated. The importance of a thorough site investigation to reduce risk of the unknown can not be overemphasised.

5. **Planning fees**   These are the fees involved in making a planning application and securing consent for the development project. Many developments necessitate a change in the use of the property, such as residential to commercial, where the highest and best use of the property has changed over an extended time period. This component normally only includes the fees paid to the local planning authority, which are based on the scale and nature of the scheme. A list of such charges can be obtained from the local planning authority.

In the above example, planning consent has already been obtained. However, in a situation when obtaining planning permission may prove difficult (especially where there is a substantial change in use), a developer has to allow for planning consultant fees and, in the event of an appeal, additional costs such as fees for solicitors, counsel and expert witnesses. The extra time period involved will need to be reflected in the interest costs associated with the extended holding period.

6. **Building regulation fees**   These are on a sliding scale based on the final building cost. Details of such fees are available from the building control department of the relevant local authority.

7. **Funding fees**   These fees are related to the costs associated with arranging development finance and will vary on the method of finance. To illustrate this point, Example 3.1 is bank financed so the developer will need to pay the bank's arrangement fees, solicitor's fee and surveyor's fee. These fees are a matter of negotiation but usually reflect the size of the required loan and may be anything between 3 and 10 per cent of the value of the loan (see Chapter 4 for details).

If the development is to be forward-funded with a financial institution, then the developer will pay the fund's agent (if appointed) and solicitor's fees, as well as their own agent (if appointed) and their own solicitor. The developer may also have to pay the fund's building surveyor's fees to monitor the construction of the building on behalf of their client.

8. **Finance costs/interest** Interest costs are a critical element of the appraisal and can have an adverse effect on the overall viability of any development proposal. These costs reflect either the actual cost to the developer of borrowing money or the implied or notional opportunity cost (reflecting the investment foregone, i.e. the capital could be earning money elsewhere at a comparatively higher return but not necessarily with the added risk). The actual cost of the finance/interest is affected by many factors including the loan-to-value ratio (LVR), the risk in this land use sector, the established relationship between the borrower and the financier, and the borrower's estimated risk that the borrowed funds will be paid in full by the due date. In Example 3.1 the development company will be borrowing money from the bank and, as a condition of the loan, will be providing some capital from its own resources. It is assumed that the interest rate charged by the bank and the opportunity cost of the developer's own money are the same.

In order to calculate the interest costs, the developer must estimate the total length of the development, at which point cash expenditure will cease and then cash inflows will occur. Normally the cash inflow will be either when the building is let to a tenant/s (and therefore becomes income-producing) or sold – the decision about either option depends on whether the developer wishes to retain the scheme or not. In addition, the developer must allow sufficient time for all the preparation work needed after the site has been acquired but prior to the commencement of the building contract. Also there must be a careful estimation of the letting-up or selling period (void period), which is based on a judgement of prevailing market conditions at that future point in time.

In Example 3.1 the development timetable is assumed as follows:

Site acquisition, preparation and pre-contract	6 months
Building contract	12 months
Letting period	6 months from completion
Investment sale period	6 months from letting
Total development period	**30 months**

These periods can be expressed in a timeline diagram (Figure 3.1).

The site acquisition is the first commitment that requires a major capital outlay and, therefore, interest is calculated on all site acquisition costs over the entire development period – this is the longest time span within the development period and runs from the date of acquisition to the eventual

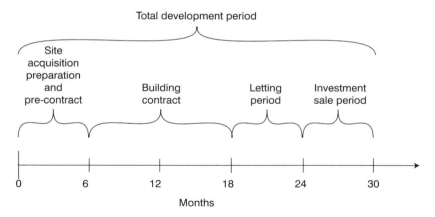

Figure 3.1   Development timeline

letting/sale of the building (i.e. in Example 3.1 it is thirty months). In some circumstance there are expenses incurred prior to the site acquisition, such as the costs associated with searching for potential sites, although such costs are usually not considered to be substantial capital outlays that attract interest. Once the building contract is signed then most of all the other costs will be incurred at various different times over the building contract period (in Example 3.1 it is twelve months), although the actual timing of these cash flows will vary and are often difficult to accurately quantify. Accordingly, a 'rule of thumb' method of calculating the interest is adopted that assumes that costs are incurred evenly over the contract period. Therefore, all the costs, with the exception of promotion and letting costs, are divided in half and then the interest is calculated on that sum over the whole period. In Example 3.1, the time period is twelve months.

Once the building contract is complete, then the interest will continue to accrue on all the building and other costs spent (except some of the promotion costs and letting/sale fees) until the date when it is assumed that the building will be let/sold. In Example 3.1, it is assumed that the building is let within six months of completion and then sold to an investor within twelve months of completion. It is further assumed that six months rent-free are granted to the tenant. If it were to be assumed that some rental income would be received before the sale of the investment then such income would be included in the appraisal and offset against the interest calculation. In Example 3.1, interest is calculated by using the Amount of £1 formula for compound interest (Baum and Mackmin, 1989). In order to calculate compound interest on a quarterly basis the interest rate of 8.25 per cent is divided by 4 to obtain the quarterly rate of 2.06 per cent, which produces a compound interest formula of $(1.0206)n$, where 'n' represents the number of quarters over which the interest is calculated.

9. **Letting agent's fees**   These fees relate to the cost of the agent letting the building to new tenants – the actual amount will vary depending on factors such as the number of letting agents competing for the letting rights (e.g. the profile of the development in the marketplace), as well as the demand by tenants to rent space in the development. If joint agents are involved these fees are usually 15 per cent of the rental value achieved at letting. If only one agent is involved, then the fee is reduced to 10 per cent of the rental value achieved at letting. In some circumstances, the developer may negotiate a fee with the agent using an incentive basis. It is usual for the tenant to pay the developer's legal fees relating to the completion of the lease documentation.

10. **Promotion costs**   The developer has to make an assessment of the likely sum of money that needs to be spent on promoting the project in order to let the property, and very often it is this element of the evaluation that is underestimated at this initial stage (see Chapter 8 on Promotion). This amount will be affected by the perceived level of demand for the development (i.e. a high-profile development may be in high demand and require less promotion) and the location of the prospective tenants (i.e. advertising in a national newspaper or via other costly mediums such as television or radio).

11. **Sale costs**   Sale fees may need to be included only if the developer intends to sell the building once it is fully let. These will include any agent's fees together with those of the developer's solicitor, representing between 1 per cent and 2 per cent of the net development value (NDV).

12. **Other development costs**   The inclusion of other costs within the evaluation will depend on the nature of the development and will be specific to the project (e.g. party wall agreements, planning agreements with the local planning authority and rights of light agreements).

If the developer considers that there may be a void period between completion of the development and letting the property, then costs such as maintenance and insurance will need to be included. If a lengthy void period is anticipated, an allowance will need to be made for additional costs such as business rates (the actual amount varies and in 2007, 50 per cent of the full rate was payable after three months on empty properties once assessed) and maintenance/management costs. In addition, if the scheme has been forward-funded by a financial institution, then, under the terms of the funding agreement, rent may be payable to the fund until a letting is achieved (see Chapter 4).

13. **Contingency allowance** Relatively few property developments, if any, are completed exactly as originally planed nor are they likely to adhere entirely to the initial budget forecast. It is important, therefore, to incorporate a contingency allowance to cover unexpected costs. However, the actual contingency itself will vary from project to project and depends on the risk profile of the developer, the developer's ability to plan and execute an accurate development plan, the associated time period, the level of risk/return built into the proposal and the level of flexibility. In Example 3.1, a contingency allowance of 1 per cent was adopted.

14. **Developer's profit/risk allowance** The residual in this appraisal is the developer's profit/risk allowance, which is usually expressed as a percentage of the total development costs or, alternatively, as a percentage of the NDV. As per standard economic theory, the profit that a developer will require will depend on the degree of risk involved with the scheme – a higher level of risk will be commensurate with a higher level of return and visa versa. It is difficult to generalise but often developers will seek between 15 per cent and 25 per cent of the total cost; the percentage rising with the perceived risk (see 'Identifying uncertainty and risk'). The profit may contain an element for contingencies (rather than a separate allowance for contingencies as discussed above).

If the developer is an investor wishing to retain the development, then profit may be assessed by reference to the yield on cost (in Example 3.1 it is 8.60 per cent). The yield or return on cost is the total development cost (excluding profit) divided into the first year's rental income. The resulting yield needs to be higher than the yield applied to obtain the NDV (which is comparable to the yields on similar standing investments) as the difference between the two yields represents the profit to the investor.

In Example 3.1, the land price used in the evaluation is the asking price and, therefore, is fixed. However, in most cases the developer has to establish the land price that can be afforded in order to enjoy a fixed target rate of profit. At the same time, the landowner (vendor) will be seeking to maximise the sale price, and may not even quote an asking price. In Example 3.2, we assume the developer wishes to ensure a rate of profit (risk/return) of 20 per cent on total development costs, so a residual land evaluation is carried out to determine the affordable land price.

Example 3.2   Residual valuation

	£	£
(a) Net Development Value (NDV)		
(i) Estimated Rental Value (ERV)	1,032,573	
Net lettable area 38,250 sq ft (3,553 sq m) @ £27 p.S.f. (£290.62 p.s.m.)	14.29	
(ii) Capitalised @ 7.0% YP in perpetuity	14,751,041	
(iii) less purchaser's costs @ 2.75%	405,654	
Net Development Value (NDV)		14,345,387
(b) Development costs		
(c) Building costs		
Estimated building cost		
Gross area 45,000 sq ft (4,181 sq m) @ £120 p.s.f. (£1291.66 p.s.m.)	5,400.000	5,400.000
(d) Professional fees		
Architect @ 5%	270,000	
Structural engineer @ 2%	108,000	
Quantity surveyor @ 2%	81,000	
M & E engineer© 1.5%	108,000	
Project manager @ 2%	108,000	
		675,000
(e) Other costs		
Site investigations - say	17,500	
Planning fees - say	6,000	
Building regulations	30,000	
		53,500
(f) Funding fees		
Bank's legal/professional fees - say	60,000	
Bank's arrangement fee	90,000	
Developer's legal fees - say	50,000	
		200,000
(g) Finance costs		
(i) Interest on (c) building costs, (d) professional fees, (e) other costs and (f) funding fees divided by a half (= £6.33m/2) over building period of 12 months @ 8.25% compounded quarterly = $(1.0206)4$	269,238	
(ii) Interest on (c) building costs, (d) professional fees, (e) other costs and (f) funding fees (£6.33m) over void period of 12 months @ 8.25% compounded quarterly = $(1.0206)4$	538,477	
		807,715

continued…

Example 3.2   continued

(h) Letting and sale costs

Letting agents @ 15% ERV	154,886	
Promotion	95,000	
Developer's sale fees @ 1.5% NDV	215,181	
		465,067

Net total development costs excluding land costs and interest on land costs ... 7,601,282

(i) Developer's profit

@ 20% on net total development costs(£7,744,736) excluding land costs and interest on land costs ... 1,520,256

(j) Net Total Development Costs (NTDC) ... 9,121,539

(k) Residue i.e. NVC less NTDC ... 5.223,849

This residue is made up of the following elements:

Land price =	1
plus cost of acquisition @2.5%	0.025
	1.025
multiplied by cost of interest of holding land for development period and void (30 months) @ 8.25% compounded quarterly (1.0206)10 =	1.226
Total land cost	1.257
multiplied by profit on total land cost @ target rate of 20%	1.2
	1.508

The residual land value i. e. the price the developer can afford to pay to ensure the target rate of profit, is therefore derived as follows:

Residue	5,223,849
Divided by factor (calculated above) to take account of land price, acquisition costs, interest and profit as calculated above	1.508
Residual land value	3,464,136
Say	3,464,000

This calculation can be checked as follows:

Land price	3,464,000
plus cost of acquisition @ 2.5%	86,600
Total land costs	3,550.600

continued...

Example 3.2   continued

multiplied by interest for 30 months @ 8.25% compounded quarterly = (1.0206)10 = 1.226	804.155
Total Land Cost (TLC)	4,354,755
plus Net Total Development Cost excluding profit	7,601,282
Total Development Cost (TDC)	11,956,038
Net Development Value, as above	14,345,387
Less Total Development Cost (TDC)	11,956,038
Developer's profit	2,389,350
Developer's profit on cost	20.0%

Note: this result confirms that at a land price of £3,464,000 the target level of profit of 20% can reasonably be expected to be achieved.

Discussion point

What are the various components of value when undertaking a property development?

Cash flow method

The conventional method of evaluating a proposed property development, as shown in Examples 3.1 and 3.2, has two basic weaknesses. Firstly, it is inflexible in its handling of the timing of when the expenditure and revenue actually occur. As a result, the calculation of interest costs is very inaccurate and may vary substantially – in other words, unless the projected time period is the exact same length in reality, then the evaluation is incorrect. Secondly, by relying on single-figure 'best estimates', it hides the uncertainty that lies behind the calculation.

The first of these problems can be overcome by carrying out a cash flow appraisal or evaluation, which enables the flow of expenditure and revenue to be spread over the period of the development – this is a model that presents a more realistic and accurate assessment of development costs and income against the variable of time. As commonly accepted, the amount of compound interest accrued over an extended period of time can have an adverse effect due to the time value of money. Therefore, the conventional evaluation shown in Example 3.1 is presented as a cash flow appraisal in Example 3.3.

As this example shows, by enabling the expenditure to be allocated more accurately over varying timelines a better assessment can be made of interest costs. The 'rule-of-thumb' conventional evaluation, described above, assumed that building costs would be spread in this way. In practice, building and other development costs are seldom spread evenly over the period. In Example 3.3 some of the development costs are incurred before or at the start of the building contract period, e.g. funding fees and some of the professional fees. Often the majority of professional fees are incurred during the pre-contract stage and early in the building contract period, as most of the design and costing work is carried out then. In Example 3.3, only 40 per cent of the building cost has been incurred after six months of the contract, which is the half-way point. The building costs in fact follow a normal S-curve irregular pattern of expenditure as follows:

Months	1	2	3	4	5	6	7	8	9	10	11	12
% Total Costs	3	10	14	22	31	40	48	60	73	85	93	97

The remaining 3 per cent of the costs represents the standard practice of holding a retention sum under the building contract, usually for a period of six months. The retention sum and period may vary and is often perceived as 'insurance' on the overall property development process.

In practice, the quantity surveyor should be consulted to assess the timing of building costs. Computer programs are available to calculate the S-curve for a particular project and convert that into the expenditure flow. The project manager can assist in assessing the flow of other costs directly related to the building costs. The timing of all other costs should be capable of assessment by the developer based on experience.

The cash flow method enables the developer to allow for such an irregular pattern of cost, giving a more explicit presentation of the flow of expenditure and a more accurate assessment of the cost of interest. It is the nature of property development that the timing of cash flows is irregular and uneven – for example, a capital outlay for a parcel of land is usually unavoidable as, in most cases, a building cannot be constructed unless the land is owned outright. In Example 3.3, the total interest figure (£1,667,000) is higher than calculated in the conventional evaluation (£1,620,228) in Example 3.1. However, given a different pattern of expenditure, with professional fees and funding fees being incurred later on, then the total interest figure may well have been lower than in Example 3.1. It is impossible to generalise or include additional assumptions, which is why the conventional 'rule of thumb' method is simple but at the same time relatively so inaccurate. In the cash flow example, interest is calculated on the outstanding balance

Example 3.3 Cash flow approach

Monthly Interest Rate 0.690%

Cost (£000)	1	2	3	4	5	6	7	8	9	10	11	12	13	14	15	16	17	18	19	20	21	22	23	24	25	26	27	28	29	30	Total
Months																															
Land cost	3,500																														3,500
Building cost						162		378	216	432	486	486	432	646	702	648	432	216													5,400
Professional fees				38		38		60	30	30	60	75	75	38	38	38	38	30													675
Other fees	9	8		6																											53.5
Funding fees						140	8	60	12		8	4							30	30											200
Letting fees																				25	25			155							155
Promotion														120	129	120	83	44						15							95
Sale fees																				11	6	30		63						215	215
VAT paid	613			8		38	39	77	45	81	98	99	89	120	129	120	83	44	6	11	6	6	0	6	0	0	0	0	0	40.5	1,813
VAT reclaimed	0	0		-613	-1.5	0	-7.5	0	-37.5	-39	-76.5	-45	-81	-97.5	-99	-88.5	-120	-129	-120	-82.5	-43.5	-6	-10.5	-6	-6	0	-63	0	0	-40.5	-1,813
(a) Sub-total (Month)	4122	9	0	-662	-1.5	215	261	514.5	265.5	504	574.5	619	514.5	708	769.5	717	432	160.5	-84	-17	-12.5	30	-10.5	419	-6	0	-63	0	0	215	10,294
(b) Balance B/F	0	4,150	4,187	4,216	3,679	3,702	3,944	4,233	4,780	5,080	5,622	6,239	6,904	7,469	8,233	9,063	9,847	10,349	10,581	10,568	10,623	10,683	10,795	10,848	11,344	11,415	11,493	11,507	11,596	11,664	11,960
(c) Total (a + b)	4,122	4,159	4,187	3,554	3,677	3,917	4,205	4,748	5,046	5,584	6,196	6,858	7,419	8,177	9,002	9,780	10,279	10,509	10,497	10,551	10,610	10,713	10,775	11,267	11,339	11,415	11,430	11,507	11,596	11,879	
(d) Interest	28	28	28	25	25	27	29	32	34	38	42	47	50	56	61	67	70	71	71	72	72	73	73	77	77	78	78	78	79	81	1,667
Balance C/F (c + d)	4,150	4,187	4,216	3,679	3,702	3,944	4,233	4,780	5,080	5,622	6,239	6,904	7,469	8,233	9,063	9,847	10,349	10,581	10,568	10,623	10,683	10,795	10,848	11,344	11,415	11,493	11,507	11,596	11,664	11,960	

(including interest) at the end of each month at the rate of 0.68 per cent per month, $\sqrt[12]{.00847}$, in order to equate to the effective annual rate in Example 3.1 of 8.47 per cent per annum.

In the above example, the project is a single office development and, therefore, cash inflows (in the form of sale or rent) would not generally occur until the entire building is complete. The advantages of the cash flow method are more clearly demonstrated in relation to developments where receipts (or cash inflows) occur during the development period prior to final completion of the entire scheme, e.g. a development of phased industrial units, a major retail scheme and a residential scheme. Another example would be a large and complex mixed-use development scheme that would take a number of years to complete fully and, therefore, would be developed in phases. In this case, it may be possible to let or sell the early phases whilst construction continues. As with most businesses, cash flow is critical due to the cost of borrowing funds and the effect of compound interest over an extended period of time – the potential to develop a property in phases can be a major advantage for the overall viability of the project. Recent developments in construction technology have assisted more building types to be developed and released in phases. For example, some offices in high-rise buildings can be let or even sold off and allow the new owners to occupy the lower floors, even though the upper floors or other sections of the building are still under construction – this example applies to large-scale projects that take years to complete, where the developer has been creative in their project management when it came to the desire to commence cash inflows at the earliest available opportunity.

There are other advantages of adopting the cash flow method. For example, the model enables the developer to adjust for changes in interest rates easily over the development period or for different sources of finance within the appraisal. In addition, this method disciplines the developer to think hard about the nature of the cash flow of the project. It highlights, where possible, the need to delay outgoing payments and bring forward receipts (cash income). It shows the developer that cash flow is an important tool in identifying a competitive advantage over competitors, which can be achieved by maximising profitability and reducing the cost of borrowing. A developer will certainly have to produce a cash flow appraisal to satisfy potential sources of finance, when a detailed business case is a standard request prior to funds being advanced. In reality, many developers use both conventional and cash flow techniques. They will use the cash flow method to calculate the interest cost and input the resultant figure into a conventional evaluation for presentational purposes. In addition, the cash flow method will be used throughout the development period to constantly evaluate the project as costs are incurred and influencing variables (e.g. interest rates) change.

It is important for the developer to assess the impact of taxation on the project, which may involve many different charges for different government authorities and is constantly subject to change. In the UK the issue of tax and VAT remains a relatively complex issue, and there usually are VAT implications to be factored into the development appraisal. For example, VAT is chargeable on commercial property construction costs and rents but companies only pay net tax to the government so VAT on costs can be offset against VAT on sales. Therefore, different types of developers and different types of schemes all have different VAT implications including standard, reduced and zero-rated VAT scenarios.

VAT has an impact on the cash flow of a development project. Although, in the majority of cases, developers will be able to fully recover the VAT paid on land transactions and construction costs if they elect to charge VAT on the sale of the completed building or on rents from the letting of the completed building, a cash flow implication arises as there may be a delay between the payment of VAT and its recovery. In Example 3.3, a three-month delay is assumed. In this case, the delay has no effect on the overall interest figure but, on larger schemes, the delay in repayments will almost certainly impact on the interest calculation due to the leverage involved. The legislation on VAT is very complex and it is beyond the scope of this book to examine all of the implications in detail. Nevertheless, a property developer must be fluent and up to date with any taxation implications or government restrictions relating to money. It is important to stress that a developer must fully assess VAT and its direct or indirect effect on a particular development project when carrying out an appraisal. In particular, VAT may affect the rent the developer may achieve on completion of the project if the likely tenant is in an exempt business (i.e. they are unable to recover VAT), e.g. banks, building societies or insurance companies.

Discounted cash flow methods

Although a discounted cash flow (DCF) also examines the different cash flows, they are all discounted back (using a present value formula) to a common point in time to facilitate an even comparison or analysis. The discounting component acknowledges the relationship between time and money, which is especially relevant in property development – for example, the extended period between when the land is purchased and when the building is completed and cash inflow commences. Example 3.3 calculates interest on a month-by-month basis to reflect a normal development pattern, so that at any point in the development programme the developer can establish the outstanding debt at that particular time. The time periods can be modified to any time period, such as days or years, depending on the intended complexity of the DCF. Alternatively, two other cash flow techniques can be used, known as the 'net terminal approach' and 'discounted cash flow' (DCF) methods. As Example 3.4 shows, the net terminal approach simply calculates the interest in a different way but produces exactly the same result as the normal cash flow method. The interest is calculated on each month's total expenditure until the end of the development period, i.e. when the development is let (or sold) and the debt is fully repaid (plus profit/risk allowance). Note that the net terminal approach will overstate the amount of debt outstanding at the end of each month and has no advantage to the developer over the normal cash flow in Example 3.3. The model displayed in Example 3.3 is in the format of a traditional cash flow where the time periods are on the X axis and the variables for each time period are listed on the Y axis.

The DCF method is distinctly different as it does not calculate interest on the monthly expenditure. Instead, it sums the income and expenses for every month and then discounts the amount for each month back to present day equivalents to establish the value of the profit in today's value (rather than at the end of the development). The discount rate used is the cost of borrowing the money and the formula used to convert costs and values to the present day is the 'Present Value of £1', which is $1/(1+i)n$ (or alternatively $(1+i)^{-n}$), where i represents the prevailing interest rate (e.g. 0.075 for 7.5 per cent) and n represents the number of periods (e.g. in months). This formula is the reciprocal of the amount of £1 used for compound interest.

Using the same figures as those contained in Example 3.3 a DCF is calculated in Example 3.5.

The main advantage of this approach to the developer is that it allows a subsequent calculation of the 'internal rate of return' (IRR), which is the measure used by some developers to assess the profitability of a scheme since IRR considers both the timing of the cash flows and the magnitude of each cash flow. This is as opposed to examining just a percentage return on cost (without consideration to the timing of the cash flows) or the present value of the profit (which doesn't fully consider the initial financial outlay and the degree of risk the developer is exposed to). Therefore, the DCF method is more likely to be used by investors who wish to retain the development within their portfolio and also seek to analyse the return on their investment. In order to calculate the IRR, the discount rate is varied by trial and error to the rate that will discount all the future costs and income back to a present value of zero. In other words, this is the percentage return when the project does not make or lose any money from the initial outlay. The IRR is also

Example 3.4   Net terminal approach

Months	Cash-flow (£000)	Interest until completion at 0.68%	Total (£000)
1	4,122	1.2255	5,052
2	9	1.2172	11
3	0	1.2090	0
4	-562	1.2008	-675
5	-1.5	1.1927	-2
6	215	1.1846	255
7	261	1.1766	308
8	514.5	1.1687	602
9	265.5	1.1608	309
10	504	1.1529	582
11	574.5	1.1452	658
12	615	1.1374	700
13	514.5	1.1297	582
14	708	1.1221	795
15	769.5	1.1145	858
16	717	1.1070	794
17	432	1.0995	475
18	160.5	1.0921	176
19	-84	1.0847	-92
20	-17	1.0774	-19
21	-12.5	1.0701	-14
22	30	1.0629	32
23	-10.5	1.0557	-12
24	419	1.0486	440
25	-6	1.0415	-7
26	0	1.0345	0
27	-63	1.0275	-65
28	0	1.0205	0
29	0	1.0136	0
30	215	1.0068	217

Total Development Cost	11,960
Net Development Value	14,345
Profit	2,385

Example 3.5   Discounted cash-flow approach

Months	Cash-flow (£000)	Interest until completion at 0.68%	Total (£000)
1	4,122	0.9932	4,094
2	9	0.9865	9
3	0	0.9799	0
4	-562	0.9733	-547
5	-1.5	0.9667	-1
6	215	0.9602	206
7	261	0.9537	249
8	514.5	0.9472	487
9	265.5	0.9408	250
10	504	0.9345	471
11	574.5	0.9282	533
12	615	0.9219	567
13	514.5	0.9157	471
14	708	0.9095	644
15	769.5	0.9033	695
16	717	0.8972	643
17	432	0.8912	385
18	160.5	0.8852	142
19	-84	0.8792	-74
20	-17	0.8732	-15
21	-12.5	0.8673	-11
22	30	0.8615	26
23	-10.5	0.8557	-9
24	419	0.8499	356
25	-6	0.8442	-5
26	0	0.8384	0
27	-63	0.8328	-52
28	0	0.8272	0
29	0	0.8216	0
30	215	0.8160	175
31	-14345	0.8105	-11,627

Net Present Value (profit)			-1.937
Net Present Value with interest @ 0.068% for 31 months =			2.390

ideal for comparing different potential property developments with their own variations in the timing and size of the cash flows, e.g. comparing a small residential development with a large multi-storey office building. However, the disadvantages of this method are that the DCF method does not show the outstanding debt at a particular time and the profit in today's value rather than the actual sum that will be received at the end of the development.

Identifying uncertainty and risk

Although the cash flow method provides a more accurate and explicit form of calculation, it still relies upon a set of fixed variables. That is, the elements that make up the calculation, such as building cost and rent, are presented as selected 'best estimates' without giving a true impression of the range from which they have been selected. If we look more closely at the basic example of a conventional evaluation set out in Example 3.1 above, we can see that it is based on a considerable number of variable factors, as follows:

1. land costs
2. rental value
3. square footage (or metres) of building
4. investment yield
5. building cost
6. professional fees
7. time – pre-building contract, building and letting/sale periods
8. short-term rates of interest
9. agents' fees
10. promotion costs
11. other development costs.

Since, in this example, the land purchase price is fixed, these eleven variables can be reduced to the four main groupings listed below, which will most affect the profitability of a development project:

1. short-term rates of interest
2. building cost
3. rental value
4. investment yield.

It is important that the financial information input into the cash flow model is as reliable as possible, and the level of reliability depends on the experience

and sources of information the developer uses. Recer
completed by a property developer are often a good startin
allowances must be made for changes in supply levels a
occurred since. In addition, developers normally rely to a
the professional advice of the development team to estimate the cost of
main variable groupings outlined above. The quantity surveyor and project
manager advise on building costs and related costs. The agent will advise on
rental value and investment yield, and hence the likely development value.
However, in the end, developers must form their own judgement about the
estimates that are made of the variable factors. They have to assess the likely
risk of the main variables changing when deciding on the required level of
return in the evaluation process. It is important that the developer uses up-
to-date rental values and building costs to reflect income and expenses in
every development appraisal. Due to the complexity of the property market
and the interaction of many variables and influencing factors, valuers are
unable to predict future changes in property and rental values. Accordingly
it would not be advisable for a developer to predict future rental values, even
when building costs in the appraisal are inflated at current inflation rates, as
this would expose the developer to more risk. It cannot always be assumed
that rises in building costs during a development, which are common and
standard practice due to rises in inflation levels, will be saved by rises in
values.

The rental income, investment yield and building costs are usually the
most sensitive variables and commonly are subject to external fluctuations
outside the control of the property developer. To fix the level of rent, the
developer may be able to secure a pre-let commitment with a tenant, and
to fix the investment yield it may be possible to pre-fund the scheme with
an appropriate institutional investor. Either or both of these options may
be achieved either before or during the development project. In this way,
rather than the developer placing the emphasis on the financial risk with
the property development, they can now focus on the project management
aspects such as ensuring the project is built within budget and on time.
Much of this will depend upon the quality of project management, but in
some cases a fixed-price building contract may be secured. Both of these
alternatives are discussed further below and it must be acknowledged that by
reducing or effectively sharing the risk, the developer must expect to have to
limit the potential reward.

Risk is embedded throughout the property market and is the starting point
for any analysis involving property. The two major types of risk that affect
property are either systematic (or market) risk or unsystematic (or property-
specific) risk. Importantly, a developer should never underestimate risk and

the level of risk in every development scheme should be identified and, if possible, contained or reduced. The next sections will now examine ways in which some of the variables may be fixed in order to limit the developer's exposure to risk. It is important to remember that as the development process progresses, the developer's commitment increases and the possibility of variation decreases, and these both equate to a higher degree of uncertainty and associated risk.

Land cost

As previously discussed in Chapter 2, the purchase price of the land (either vacant or with an existing older improvement) is usually the first main financial commitment. In order to reduce risk, a site should not be purchased until the appropriate planning permission has been obtained and the detailed building cost established. If this is not possible, the developer should try to negotiate a contract that is subject to the obtaining of a satisfactory planning consent and this is standard procedure for many property sales. If the outcome of the planning application is uncertain at the date of agreement, then it may be possible to negotiate an option to purchase the land by a future date once planning permission has been obtained. Alternatively, a joint venture arrangement might be entered into with the landowner whereby the land value plus any accumulated additional 'notional' interest might be calculated at a future date during the development period.

Once the land is purchased the developer is committed to a particular location, which cannot be changed and which in turn will have a major influence on the highest and best use of the land. In other words, the value of the land and any development scheme built upon it might be affected by external physical factors such as a new road. Depending on market conditions the developer may be able to make a profit by simply selling the land prior to the commencement of the development scheme. Once planning consent has been obtained, the value of the scheme is established, although further applications may be made to improve the value of the site. However, planning applications take time and any improvement in value that might be obtained needs to be balanced against the costs of holding the site.

Building cost

The building cost is the second major financial commitment or capital cost and a number of other costs relate directly to its final sum, e.g. professional fees. Once the building contract is signed the developer is committed to a certain cost, which invariably will move upwards (and rarely downwards), not least due to the effect of inflation over the construction time period. In addition, many cost increases experienced during the development period are due to the developer's variations or late production of information by the professionals responsible for design. These are matters over which the developer must exercise tight control. There are some ways of making the building cost more certain by passing all or some of the risk and design responsibility onto the building contractor, although greater certainty of cost usually means a higher building cost. These are described in greater detail in Chapter 6. Project control is of vital importance in preventing increases in both building cost and time delays, as well as decreasing the risk that the builder will pay penalty rates for a late handover (i.e. substantially longer than the agreed contract date). Therefore in many instances, the employment of an experienced project manager is advisable. Furthermore, it is important that the developer and/or project manager constantly questions every aspect of the building contract in order to contain any problems as they arise.

Rental value

It is essential to obtain the most reliable, up-to-date estimate of rental value. Due to the relatively large size and, therefore, lettable area of some property developments, an error in the rental estimate on a rate per square metre basis can have a significant adverse effect on the estimated aggregate income. A reliable estimate of rental value must be undertaken via a thorough analysis of the prevailing market as well as in consultation with the letting agents (if they are to be appointed). However, the level of uncertainty associated with achieving an estimated level of rent can be removed if a pre-letting by tenants can be achieved. Due to nature of the property market, when the development is actually completed there may be an oversupply and the property may be difficult to let. Accordingly, some developers might not proceed with a particular development until a pre-letting is achieved to reduce a considerable element of the risk involved. For example, with a business park scheme or a large industrial scheme, the developer may provide all the necessary infrastructure and landscaping initially and then build each element on a pre-let or a 'design and build' basis. In addition, the developer may build one or two speculative units to show potential occupiers the type of building that could be provided and adapted to suit their individual requirements. Quite often the developers of major shopping schemes need to secure the major tenants to the large units (referred to as 'anchor' tenants) at an early stage in order to attract retailers to the smaller 'unit' shops. Financiers are also reluctant to commit to lending money unless there has been a major pre-commitment.

Nevertheless, the benefit of achieving a pre-letting and thereby reducing risk has to be weighed against the opportunity costs of achieving a potentially higher profit in a rising market. In the time it takes to complete the development scheme, rents might rise; alternatively, there may be more tenancy demand when the development is nearing completion and a prospective lessee can actually visualise the lettable area, rather than just looking at the architectural drawing. In the case of anchor tenants in a shopping scheme, the developer may have to pay what is called a 'reverse premium' to the retailer to secure a pre-letting and ensure the overall property development will receive approval from the financier. Of course, the cost of such a premium must be accounted for in the development appraisal. An additional advantage of securing a pre-letting is to reduce the overall development timetable before income is received, as the building will be handed over on completion without the uncertainty of a void period and further interest payments.

Short-term interest rates

Unless the scheme is being financed entirely by the developer, funding arrangements need to be in place before any major commitment is made. In obtaining the necessary finance to acquire the land and build the scheme, the developer will be exposed to any fluctuations in short-term interest rates. However, at a cost, the developer has the option to either fix or cap the interest rate (see Chapter 4). If the developer achieves a forward-funding of the scheme with a financial institution then the interest rate agreed with them may be fixed.

Investment yield

Investment yield is determined by the property investment market. This is the relationship between the total value of the property/improvements and the total rent received. It should be noted that this relationship varies at any particular point in time according to market factors such as the supply of competing developments, investor demand and rental growth. However, the uncertainty of the yield changing over the period of the development can be removed if the scheme is pre-sold or pre-funded. If a scheme is pre-sold to an owner–occupier then the developer is really performing the role of project manager. With a pre-funding the developer secures both short-term and long-term finance by agreeing to sell the completed and let scheme to the financial institution. Although the developer still bears the risk of securing an acceptable tenant on satisfactory terms and controlling building costs, as with pre-letting, the terms negotiated prior to the commencement of the scheme are likely to be less favourable to the developer than those that can be negotiated at the end of the project. The developer's chances of securing pre-funding improve if a pre-letting is in place.

When a developer is deciding on how much to reduce the element of risk, a balance needs to be struck between profit and certainty. In general terms, the greater the certainty, the lower the potential profit. The level of risk a developer is prepared to accept will depend largely on their motivation. Occupiers, contractors, financial investors and the public sector involved in the property development process will all be looking to reduce risk to an absolute minimum, although development companies may be willing to accept a much greater degree of risk in return for higher rewards. The degree of risk is usually directly related to the complexity and scale of the proposed development. For example, at one extreme a small, self-contained office block pre-let to a major corporation represents a very limited degree of exposure to risk by the developer. At the other extreme a substantial degree of risk is involved in assembling, over a long period of time, a large town centre site suitable for a comprehensive mix of uses including shops, offices and residential. Where a high degree of risk is perceived, it is usual for developers to seek development partners in order to share both the risks and rewards.

Uncertainty is endemic in the process of appraising development opportunities and great attention needs to be given to pre-project evaluation to identify and evaluate the balance between risk and reward. The few assumptions that can be relied upon will reduce risk associated with the property, which in turn will increase the likelihood of a successful project. The cost of such work, and the time it takes, often leads to greater savings of cost and time later on in the project. However, the time available to the developer at the pre-project evaluation stage is usually limited, especially in a competitive tender. In this situation, the developer's judgement and expertise is critical. Establishing the economic viability of the scheme and the particular characteristics of the marketplace before being committed to the major financial burdens of land and building costs is most important. Only when the evaluation has been prepared and discussed by the development team can a decision be made as to whether or not it is prudent to purchase or lease a particular site and, if so, on what terms and subject to what conditions. Often a financial component of a property development, such as the initial land purchase price, may be higher than market expectations. Any additional money outlaid for the initial land purchase must be deducted from the developer's profit, which could quickly ensure the project is not viable. It is essential that individual variables in the assessment are accurate and each variable reflects current market value, as the income components will be based on current market value.

Discussion point

Why are uncertainly and risk major considerations when using the cash flow method?

Sensitivity analysis

One of the critical questions posed by the evaluation process is how a developer measures the uncertainty involved in a scheme and, therefore, how much profit is required to balance the resultant risk. Developers are often criticised for not sufficiently understanding and analysing risk. This is a valid criticism, as property developers can underestimate the level of risk and a project may not reach completion due to unforeseen problems; on the other hand, developers cannot afford to be too conservative as they may never be successful in securing sites. A careful balance has to be struck that relies entirely on the developer's judgement and experience. It is worthwhile to identify and examine methods of analysis available to assist the developer to examine the level of risk.

In the previous section, once the land price was known and fixed, the main variables of the evaluation were identified as being short-term rates of interest, building cost, rental value and investment yield. In most cases, the financial outcome of the development is more sensitive to their variability than to the variability of the other factors previously mentioned because they are the highest proportional values/costs in the evaluation. For example, a 10 per cent increase in building costs is likely to have a more significant overall impact on profitability than a similar increase on promotion costs. The name given to the procedure for testing the effect of variability is 'sensitivity analysis' and, given the nature of the property market with many variables constantly in a state of change, the assessment of a potential project must acknowledge this risk and have an in-built capacity to adapt to suit. Accordingly, one or more factors in the evaluation or appraisal can be varied and the effect on viability measured and recorded. The procedure can then be repeated and the different results compared. If, for example, we take the appraisal set out in Example 3.1, we can carry out the sensitivity analysis shown in Table 3.2.

This analysis shows that the outcome of the appraisal is most sensitive to changes in investment yield, rent and building cost. If, for the purpose of our

Table 3.2   Percentage variation in developer's profit

Variable (original value)	Original value –10%	Original value +10%
Land price (3,500,000)	+23.29%	–21.64%
Interest Rate (8.25%)	+8.69%	–8.76%
Building costs (120 per m²)	+35.05%	–33.05%
Rent (27 per m²)	–59.49%	–59.13%
Gross are and net lettable area (45,000/4181 and 38,250 ft²/3553 m²)	–25.94%	–23.00%
Professional fees (12.5%)	+3.91%	–3.86%
Investment yield (7%)	+66.65%	–54.73%
Agents fees (15%)	+0.79%	–0.79%
Promotion (90,000)	+0.48%	–0.48%
Developers sales fees (1.5%)	+1.10%	–1.10%
Funding fees (200,000)	+1.15%	–1.15%
Other costs	+0.31%	–0.31%

example, we now assume that a change in investment yield is unlikely within the timescale of the appraisal, we can concentrate upon the effect of possible variations in rent and building cost. Let us suppose that possibilities that the development team think appropriate is a range of rents of £260–320 per square metre per annum and a range of building costs of £1,000–1,600 per square metre. Remember that at this stage we are talking about possibilities and not probabilities and, therefore, that the range is likely to be rather wide. Table 3.3 can now be prepared, showing the level of developer's profit expressed as a percentage (i.e. profit as a percentage of total development value).

The total range of possible outcomes can be seen to be a developer's profit of –5.62 per cent to +50.54 per cent.

The next step for the developer is to narrow the focus of attention by concentrating on the most probable outcomes. Let us suppose that, as a result of discussion among the development team, the outer limits of the ranges of rent and building cost are excluded as being possible but unlikely. The developer can now concentrate on a narrower range of outcomes given in bold in Table 3.3.

Although this represents a substantial focus of probability, the range of possible outcomes still remains wide: a developer's profit that ranges from +10.02 per cent to +17.64 per cent gives an indication of the real

Table 3.3   The level of developer's profit expressed as a percentage (i.e. profit as a percentage of development value)

		Rents, pounds per square metre						
		260	270	280	290.62	300	310	320
Building	1000	23.21%	27.80%	32.37%	37.21%	41.48%	46.02%	50.54%
costs,	1100	17.24%	21.61%	25.97%	30.59%	34.66%	38.98%	43.30%
pounds per	1200	11.83%	16.00%	20.16%	24.57%	28.48%	32.60%	36.72%
square	1291.66	70.28%	11.29%	15.29%	19.53%	23.26%	27.24%	31.20%
metre	1400	20.36%	60.19%	10.02%	14.07%	17.64%	21.43%	25.22%
	1500	−1.79%	10.89%	50.56%	90.45%	12.88%	16.53%	20.17%
	1600	−5.62%	−2.08%	10.45%	50.19%	80.49%	12.00%	15.51%

uncertainty that lies behind the appraisal. The developer must now try to weigh up the possible outcomes, assigning either objectively or subjectively some probability to each estimate of rent and building cost. In the end, the original 'best estimate' of £290.62 per square metre per annum rent and £1,291.66 per square metre building cost may be selected, but the context of possibility and uncertainty in which it lies can now be better understood. On the other hand, an attempt may be made to fix one of the variables in one of the ways we discussed above. Let us assume, for example, that a pre-letting is agreed at £290.62 per square metre per annum. This now narrows the range of likely outcomes to +14.07 per cent to +30.59 per cent. On these figures, a maximum profit of +34.66 per cent has fallen to +30.59 per cent, but as a trade-off the minimum level of profit has risen from +10.02 per cent to +14.07 per cent and the degree of uncertainty has been reduced. It is just this kind of trade-off that is made possible by sensitivity analysis and by the understanding of probabilities, particularly when they are matched to the use of cash flow appraisals.

This is a brief introduction to the idea of sensitivity analysis based on relatively straightforward examples. In carrying out sensitivity analysis at the initial evaluation stage, the developer is weighing up the balance between risk and reward. The level of uncertainty in the project is, therefore, a most important factor. Uncertainty can be reduced by fixing any of the four variables in the ways we have discussed.

Conventional methods of evaluation do not provide any indication of the uncertainty that is an inherent part of the development process. Whilst cash flow methods of appraisal overcome the inaccuracies of the conventional approach, they still only represent a 'snapshot' of the viability of the scheme. Sensitivity analysis is a tool in the developer's decision-making process and can provide a measurement of the risk of the development scheme. It forces the developer to be more specific about the assumptions and estimates made. It assists but does not replace a balanced and informed decision-making process.

However, there is a danger in relying too heavily on the figures produced in the financial evaluation of a scheme. A developer must avoid using the evaluation process to justify a development project that on the face of it looks good – often referred to as a 'gut' feeling. Although the evaluation must be thorough and based on the best possible information, it should be approached from the point of view of what can go wrong. Even if the figures indicate a viable scheme the developer should always research the market for the proposed development in the particular location (see Chapter 7).

Reflective summary

We have examined the conventional method of evaluation used by developers to assess the profitability of a development scheme, or the land price that can be afforded, given a required return, on any particular site. It has been recognised that the conventional technique is a very crude and inaccurate method. The inaccuracy in the calculation of interest costs can be overcome by using any of the cash flow techniques including the net terminal approach and the DCF. However, all the methods discussed only produce a residual figure, based on best estimates at the date of the evaluation, which hides the true uncertainty of the outcome of the development. Sensitivity analysis and a thorough analysis of underlying market conditions improve the assessment of uncertainty and risk.

Development finance

4.1 Introduction

The majority of property developments are undertaken using funding from an external, third-party source, where the financier provides the difference between the developer's available equity or cash equivalent, and the total cost of the project including all associated expenses over the development period until completion and the return of the initial loan (plus interest and associated costs). There are two forms of finance that are required for property development: short-term finance to pay for the initial costs of production (i.e. purchase of land, construction costs, professional fees and promotion costs) and long-term finance to enable developers to repay their short-term borrowing/loan and either realise their profit via selling or retain the property as an investment. Either option depends on the developer's motivation, their financial situation and the prevailing market conditions. In this chapter we will examine the various sources of finance available to developers and the various methods of finance highlighted using worked examples.

4.2 Sources of finance

Most developments are funded by a combination of equity and finance. In these instances the lending institution assumes part of the risk associated with the development, and at the same time it also charges the developer an interest and service charge designed to be commensurate with the level of risk exposure plus an allowance for profit. As with other service providers, there is a diverse range of lenders who themselves are of varying size and expertise, but most importantly have different risk levels and specialise in certain lending projects over varying time frames. At the same time, the financial market is in a state of constant change and careful consideration must be given to which type of funding is best suited to an individual project.

In the past UK clearing banks and merchant banks have been the traditional providers of short-term development finance, with long-term investment finance being provided by the financial institutions (insurance companies and pension funds) and property investment companies. At times the financial institutions also take on the role of short-term financier by forward-funding development schemes: they provide the necessary interim development finance to a developer and agree to purchase the property on completion of the scheme. More recently there has been the emergence of Real Estate Investment Trusts (REITs), which commenced in the UK in 2007 and provided an additional source of funding for property. As a starting point it is important to briefly review the history of property financing.

Historical perspective

The role of the various financiers within the property development process has varied depending on the position of both the business and property development cycle at any particular time in relation to the credit cycle. It is important to appreciate that financiers are in the business of making money, where property is only one of a number of assets they can invest in and also lend money against. However, in general, real estate offers the financier a relatively secure form of investment since the financier can hold a mortgage or first claim over the land component including any improvements affixed to the land. The security is linked to restrictions placed on the property title, since the property owner is unable to transfer or sell the land without first clearing the mortgage on the title. In other words, the first step a prospective mortgagee undertakes is to ensure the title is clear and mortgage-free, which ensures a first mortgage has a priority claim to be repaid, even before the property owner.

Each of the various financier groups mentioned above will have different motivations and liabilities, influencing their policy towards property as either an investment or as a security against a loan. Since the 1960s developers have been able to move from one financial source to another depending on the investor/lender attitude prevailing at any particular time. Before the 1960s the roles of the short-term financiers (the banks) and the long-term financiers (mainly insurance companies then) were quite separate. Developers usually retained their completed developments as long-term property investments.

Short-term finance was typically provided by the clearing or merchant banks in the form of loans secured against the site and sometimes the buildings. Long-term funding, often pre-arranged, was generally provided by insurance companies by way of fixed-interest mortgages. On some occasions, the development would not be retained by the developer but sold as an investment to an insurance company or directly to an occupier. For example, this might happen when the developer could not arrange appropriate mortgage terms. With the exception of some merchant banks, it was unusual for the financiers to participate in the profit or risk of the development.

As inflation became a permanent feature of the British economy the insurance companies saw the disadvantages of granting fixed-interest mortgages and wanted to participate in the rental growth. At the same time, long-term interest rates rose and developers were faced with an initial shortfall of income over mortgage interest and capital repayments, often referred to as the 'reverse yield gap', which has remained an almost permanent feature of property financing (the gap narrowed and at times disappeared in the early 1990s). Thus, insurance companies grew less inclined to grant mortgages and developers were forced to give away some share of future rental growth in order to close the 'gap'. The insurance companies became more directly involved with the ownership of property and in this period the volume and balances of UK loans and mortgages decreased (see Figures 4.1 and 4.2). An increasingly active property investment market emerged and the traditional division of the roles began to blur.

Initially, to attract the best investments, long-term investors competed with and took on the additional role of the short-term financiers. Simultaneously, some of the traditionally short-term financiers, the clearing banks and the merchant banks, began to seek a share in the equity of the development itself. As the competition for the best (the 'prime' investments) increased some of the insurance companies and pension funds – either on a project basis or by the acquisition of property companies – began to take on the role of the developer, accepting the additional element of risk in return for a marginally better long-term yield. The funding of developments on a long-term basis became dependent on the property satisfying the criteria of investors. Developers had a much wider choice of financial sources.

This level of activity increased to the height of the second post-war boom of 1971–3. Then, in late 1973 and early 1974, as a result of the rise in short-term interest rates, the rent freeze and the proposals for a first-letting tax, the property boom collapsed. Although Figures 4.1 and 4.2 show that money continued to be invested by the institutions during that period, this was largely as a result of prior commitments. A more accurate reflection of the market conditions at that time is shown by the change in investment

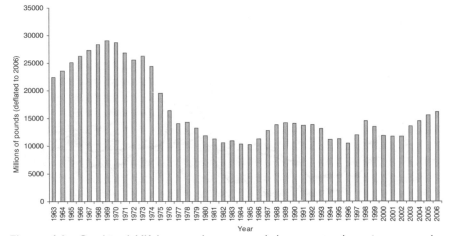

Figure 4.1 Combined UK loans and mortgage: balances – total net investment by insurance companies, pension funds and trusts (Source: Office for National Statistics, 2007b)

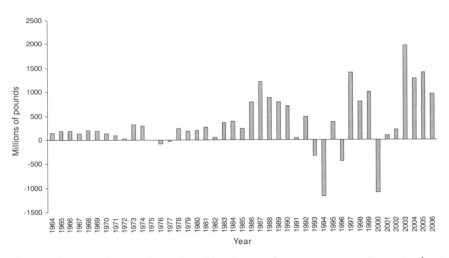

Figure 4.2 Annual change in combined UK loans and mortgages transactions – total net investment by insurance companies, pension funds and trusts (Source: Office for National Statistics, 2007b)

yields in Figure 4.3. During this period virtually no new funds were made available for development and very few developers took on new schemes.

By the end of 1977 the market for completed and let investment had re-established itself firmly with yields returning to levels of 1971. The

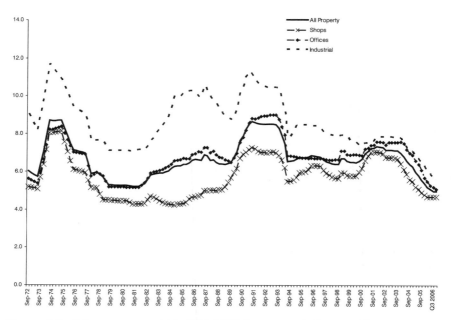

Figure 4.3 Prime property yields 1972–2006 (per quarter) (Source: CB Richard Ellis, 2007)

institutional investors emerged as the dominant force in prime commercial and industrial property markets during the first half of the 1980s. The criteria they adopted became increasingly narrow, leading to a widening of the gap between 'prime' or 'institutionally acceptable' properties and the rest. By 2006, prime investment yields had fallen to a long-term low, as is shown in Figure 4.3. Due to their selective approach, institutions kept yields low over a long period, despite rises and falls in the general level of interest rates. Many of the larger insurance companies began to carry out their own developments.

The banks became the dominant source of finance during the late-1980s development boom, encouraged by the rapid increase in rents and capital values caused by occupier demand. They replaced the institutions who reduced their property investment in the mid-1980s (see Figures 4.1 and 4.2), largely due to the better performance of other forms of investment such as equities (stocks and shares), compared with the poor performance of property in the early 1980s. Furthermore, developers preferred to obtain short-term 'debt' finance from the banks to enable them to sell their completed development into a rising market. Alternatively, developers were able to secure medium-term loans or refinance initial short-term loans to enable them to retain their developments as investments. During the boom period in particular, new financiers entered the commercial property market including foreign banks, foreign investors and to a lesser extent building societies.

The property market crashed in 1990 caused by a combination of the economic recession, high interest rates and an oversupply of new buildings. The banks emerged laden with property debt (see 'Banks and building societies') and many development companies went into receivership or concentrated on their levels of debt. As Britain's economy slowly improved after 1992–3 some development activity was gradually undertaken mainly on the basis of pre-lettings. The financial institutions re-entered the property investment market in the early 1990s, although they have never reached the benchmarks established in the late 1960s and early 1970s (see Figure 4.1). In turn this has had the effect of driving yields down (see Figure 4.4), although this graph emphasises the relatively stable level of yields over the past 20 years. This stability is due to numerous factors including increased competition for and lack of availability of 'prime' stock, a tighter monetary policy and enhanced research information available to lessen investment risk.

The twenty-first century has continued this period of sustained long-term inflation (Figure 4.5), which has had the flow-on effect of ensuring that interest rates are low in comparison to the rises in the 1980s and 1990s. At the same time there have been a variety of new financiers entering the marketplace who are keen to increase market share, coupled with the

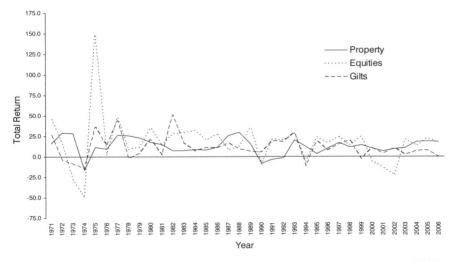

Figure 4.4 Historical sector performance – total return percentage per annum (1981–2006) (Source: IPD, 2007)

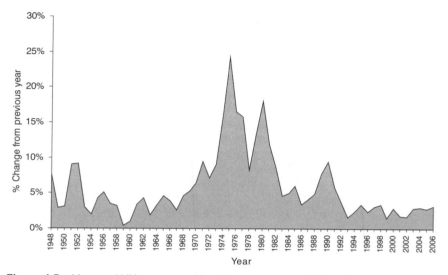

Figure 4.5 Historical UK consumer price index – percentage change from previous year (1948–2006) (Source: Office for National Statistics, 2007a)

availability of information from sources such as the internet (Dixon *et al.*, 2005). Now there is an avalanche of information available about different funding sources, as well as third-party organisations that review the attributes of each option. Whilst the global information age may offer an enormous selection of financiers who are not necessarily even based in the same country, it is important for the borrower to be fully aware of the conditions attached to any loan. For example, most financiers differ considerably in their fee structures and the developer should pay attention to the detail in the loan documents.

The increased competition has also opened up the lending market to myriad new products, which in turn have associated lending fee structures and loan lengths. Whilst the borrower has benefited from the wider choice and availability of financial products, such as the newly introduced REITs, it still remains fundamentally critical for the developer to spend adequate time reviewing the risk profile and suitability of the financier. Nevertheless, it is anticipated the rate of expansion in the lending market will slow and consolidation will occur, especially if there is an eventual downturn in the property market.

Discussion point

How have historical trends changed over time and what were the
influencing factors?

We shall now examine each of the financial sources in detail and their
influence on the property development process.

Financial institutions

A financial institution is the general term used in the property industry
to describe pension or superannuation funds, insurance companies, life
assurance companies, investment trusts and unit trusts. They invest in
property directly and indirectly through the ownership of shares in property
investment companies and development companies, which differ from
investment in REITs that are traded on the stock market. Direct property
investment includes the owning of completed and let developments, the
forward-funding of development schemes and the direct development of
sites and existing properties.

Pension funds vary considerably in size and include the individual
occupational funds managed exclusively for the employees of former/present
nationalised industries and large publicly quoted companies, together
with company and personal pension schemes managed by insurance and
life assurance companies. They invest the premiums paid by the clients to
achieve income and capital growth in real terms in order to meet the future
payment obligations of pensioners on retirement – in many ways real estate
is ideally suited as it is a secure and long-term investment. Many pension
funds and schemes are under pressure due to the approaching maturity of
their schemes (i.e. the ratio of expenditure on pensioners to income received
from premiums is increasing), which has placed the emphasis on cash flow
and real income growth. In addition, many pension schemes are linked to the
final salaries of employees, which adds to this pressure as incomes continue
to rise above the rate of inflation. Life assurance companies and insurance
companies invest the premiums they receive on life and general insurance
policies, respectively, to ensure long-term income growth to meet payment
obligations when they occur.

Unit trusts are managed by financial institutions who offer investment
management services, e.g. merchant banks. The unit trust will comprise
of unit-holders, such as small investors and institutions who are unable to
take on the risk of direct investment. The trust will manage a portfolio of

shares on behalf of the unit-holders in order to benefit from diversification and to obtain a reasonable spread of risk. There are two types of unit trust that specifically invest in property: authorised property unit trusts and unauthorised unit trusts. The former is strictly regulated to ensure that they invest in a diversified portfolio of low-risk prime income producing property as part of a balanced portfolio, as their investors include private individuals. The latter are unregulated, investing directly in a portfolio of properties and are attractive to tax exempt financial institutions such as charities and small pension funds.

The underlying goal of these financial institutions is primarily to maximise returns but minimise risk and adopt a conservative approach with every investment, since they are trustees of other people's money and, therefore, are under constant pressure to perform. In other words, the emphasis is on providing a return commensurate with a relatively low degree of risk. They invest in property as an alternative to other forms of investment such as stocks and shares (equities), together with bonds and gilts (fixed-interest income), both in the UK and globally. Note that the extent to which a financial institution will invest in property largely depends on the size of the fund and the nature of their liabilities, and will vary according to the state of the economy and the performance of property investments relative to other investments. Even though they are long-term investors they take a short-term view of performance, as they are strongly influenced by the recent performance of each type of asset, although they do forecast future trends (see Chapter 7).

It is essential to examine the advantages and disadvantages of investing in property from a financial institution's point of view. It is important to appreciate the factors that influence the investment decisions of financial institutions as this affects the funding and sale of completed development schemes.

Hedge against inflation

One of the main reasons why the institutions first entered the property investment market in the 1960s was due to the fact that property represented a 'hedge against inflation'. In other words rental growth outstripped inflation and, therefore, represented an opportunity to achieve income gains in real terms. Before the early 1990s the consumer price index was relatively volatile and often achieving double-digit growth over the previous year (see Figure 4.5). The cost of labour and constructing a new building also increased over this period, which in turn would add to the cost of a new property development. As the interest rate is linked to the prevailing inflation rate, this also added further costs onto the project. However, since the early 1990s the level of inflation has been kept relatively under control and below 5 per cent. This is due to a number of reasons, explained previously, but the tighter reign on inflation was also recognition by the government that sharp increases can have an adverse effect on the economy. Nevertheless, a property developer in a low inflation environment is more likely to have less volatility in future building costs as well as lower interest rates.

Institutional lease

A lease is an agreement enforceable by law where the property owner is guaranteed a future income at an agreed rate, usually with the ability to change the rent to reflect current market conditions. Furthermore, one of the principal advantages of real estate, as opposed to receiving cash flow from an alternative investment, is the existence of the 25-year institutional lease with upward-only rent reviews and 'privity of contract'. For example, upward-only rent reviews ensure that typically every 5 years the rental income received by the investor rises or remains at the same level and cannot fall. However, this assumes that the tenant continues to pay the rent. The English legal doctrine of 'privity of contract' ensures that in the event of a default by any tenant (unless it is the original tenant) then the landlord has the ability to require the original tenant, followed by any assignees in turn, to pay the outstanding rent. Accordingly, any tenant's contractual obligations will remain to the determination of the lease even if that tenant has assigned their interest in the lease. Tenants may assign their interest in the lease to another party provided that they have the prior approval of the landlord, which cannot be unreasonably withheld. Usually the test for such approval is the financial standing (covenant) of the tenant and typically institutions require the potential assignee to demonstrate that their last 3 year's trading profits exceed three times the rent payable. Accordingly, an investor in property is guaranteed an upward-only secure stream of income over 25 years, with some risk of voids due to tenant default.

Since the property slump in the early 1990s the prevalence of the 25-year institutional lease has been under some pressure. Notably there has been market pressure for shorter, more flexible leases by tenants in line with the requirements of their businesses. In addition, the concepts of privity of contract and upward-only rent reviews have been widely debated and criticised, due to the recent experience of tenants in the economic recession. For example when the property market is in a downturn due to its inherent cyclical nature, market rents fall below rents agreed in the height of the previous cycle, although tenants can not benefit due to the upward-only rent review provisions in their leases. In this example many tenants would go into receivership since landlords would turn to original tenants to pay the rent under the privity of contract provisions. Therefore, the government has been under some pressure from tenant pressure groups to legislate against upward-only rent reviews and privity of contract, but they have decided to allow the introduction of a code of practice to be agreed between the interest groups representing landlords and tenants.

Another reason why property has traditionally formed part of an institution's investment portfolio is diversity of performance risk. If you examine total returns from property compared to equities and gilts (see Figure 4.4) it shows that property investment is not prone to short-term fluctuations. For example, since the early 1990s property has not achieved a negative return ,in contrast to both equities and gilts. Investors are fully aware of the returns from each investment category, as well as the corresponding levels of risk.

Illiquidity and indivisibility

There are other considerations for investment in property compared with other assets, the most significant being the illiquidity of property. As property represents a large investment in financial terms it cannot be sold quickly in response to market trends, unlike equities. Selling a property can take months and it may not be possible to sell it if the market conditions are not favourable. Also the turnover of property sale transactions is relatively low compared with other assets. Selling a property usually involves high transaction costs such as agents' fees, solicitors' fees and associated government charges.

Another significant problem linked to illiquidity is the indivisibility or 'lumpiness' of a property. Usually it is not possible to sell only a small proportion of the property to increase cash flow, especially if the property development has not been completed yet. This factor alone reduces the involvement of the smaller funds in the property investment market directly. The indivisibility and illiquidity of property has become the main focus of attention for those in the industry who are trying to improve the attractiveness of property as an investment. Therefore, direct property investment of high-value real estate, such as a multi-storey office building, is usually limited to large pension funds or syndicates.

No centralised marketplace

In contrast to equities, which are traded on a centralised stock market, there is no common meeting place for buyers and sellers to transfer property. Further complications then arise as there is limited knowledge about the current state of the market, and also what the volume of trading is and the existing level of prices that are being achieved. The lack of a centralised marketplace also places pressure on the seller to pay high marketing expenses, otherwise the real estate for sale may not reach the attention of prospective buyers.

Management

Property investment is labour intensive as it also involves a high degree of management expertise measured in both time and cost. Computers and specialised programs have greatly enhanced the role of the property manager, especially with regard to monitoring payments and the current market rental levels of comparable properties at any given time. Although most management costs can be recovered from tenants under the terms of leases, it is still perceived as a disadvantage. Conversely active management of a property asset may improve the return received.

Research and performance measurement

Research into property markets has been a rapidly growing area in the twenty-first century, mainly because this specialist information is sought after by developers, investors, tenants and stakeholders. One of the main differences between property and other assets is the way in which performance is measured, making it difficult to compare property with other assets on a 'like for like' basis. The two main indicators of the state of the property market are rents and yields. The yield of a property investment is generally defined as the annual rental income received from a property expressed as a percentage of its purchase price or capital value (see Chapter 3). The yield is a measure of the property investor's perception of the future rental growth and capital growth against future risks, management expenses and illiquidity. Figure 4.3 shows the movement of 'prime yields' since 1972 with a clear distinction between all property, industrial, office and shops. Prime yields are calculated by analysis of market transactions in 'prime' properties (sometimes referred to as 'institutionally acceptable' properties), which conform to the following narrow criteria:

- modern freehold or long leasehold property;
- good location and access to services and amenities (including transport);
- highest quality and specification;
- fully let and income producing to tenants with good covenants.

However, prime properties only represent a very small percentage of the entire property investment market. The movement of 'prime' yields represents a benchmark against which the yields of all properties can be measured. The movement of yields represents market sentiment about a particular type of property and the better the perceived prospects for either capital or rental growth, the lower the yield (higher capital values relative to income). In making a decision as to how much money to allocate to property purchases the financial institutions will have regard to the performance of both the property investment market and their existing portfolio of property. There are several established performance measurement indices that measure the performance of institutional property portfolios (see Chapter 7 for further details).

As far as expectations of future performance are concerned, institutions will typically hold regular meetings to review forecasts of performance and reallocate funds to the different available asset classes. However, investment in property is not as flexible and does not sit easily with such a rapid review timetable. The integration of property into a wider, multi-asset context is generally made more difficult by the differences in terminology and valuation practice between property and other assets. This has partially been overcome by the increasing acceptance of accounting-based valuation techniques, such as the discounted cash flow approach, in order to predict the investment's internal rate of return (IRR). Previously the value of property was viewed using a relatively static approach, which in turn may affect a given level of a fund's assets held in property. It has been argued that issues related to the different terminology and valuation practices may still result in there being lower exposures to property by some fund managers than otherwise would be the case, although this resistance has gradually changed over time.

The integration of property into a wider, multi-asset investment policy is an accepted means of diversification, where a portfolio may include equities, cash and a substantial property holding. Due to the stability of property as long-term asset, many fund managers are increasingly including property in their portfolios and, therefore, comparing property directly with other asset classes.

When a decision has been made about the proportion of money to allocate to property investment, then it is the responsibility of the property fund manager to make the decision on what type of property to purchase and, importantly, in what location. Property investment policies are usually based on analysis by property type and region, looking at recent performance and future forecasts of growth (for further details on forecasting see Chapter 7). Most importantly, a major consideration is the different policies and investment criteria that each institution uses. Nevertheless, they all tend to seek a balanced portfolio of property types, although the portfolio is usually weighted towards the property type that is performing well at a particular point in time. Most also try to spread their investments geographically – for example, office investment will tend to be concentrated in London and the South East, but not exclusively. Most tend to adopt very rigid selection criteria when making decisions on what property investments to purchase. Many will only look for 'prime properties', which are openly accepted in the marketplace as at the upper end of the scale. Accordingly, they will be looking at the best located properties of the highest quality, fully let on institutionally acceptable lease terms to tenants with good covenants. However, as properties falling within the definition of 'prime property' usually account for less than 10 per cent of all properties at any one time, institutions may have to compromise on some of the following factors:

* location
* quality of specification and design
* lease terms
* tenant

Note that at times it may be impossible to obtain 'prime property' either because it is not available or the price being asked is too high (market yield is too low) relative to the perception of potential future rental growth. Therefore, some may be willing to take a balanced view on a specific property by analysing it on its own merits and adjusting the yield they are prepared to accept to reflect the additional risk. Some may give different weight to each of the above factors depending on their investment requirements. For example, those funds who are concerned with the security of income rather than capital growth prospects will put greater emphasis on the quality of the tenant covenant. The characteristics of a good location in relation to the various property types have already been examined in Chapter 2. Some institutions may be prepared to look at what may be perceived by the market as secondary locations, where there is an undersupply of quality stock, although accompanied by strong prospects of future potential rental growth.

The institutions have already adapted over recent years to shorter leases, as leases of 15 to 20 years are more consistent with the economic life of a property. If in the future the majority of tenants resist upward-only rent reviews and privity of contract provisions in their leases then the institutions will need to reappraise their approach to property investment. Some consider it will mark the end of property as an investment whilst others argue it will bring property into line with other asset types. In reality, tenants would have to pay higher rents for such flexible terms and yields would rise to reflect the risk of fluctuating incomes.

As well as purchasing completed and partially/fully let developments as investments, many institutions also carry out their own developments or provide development finance. On the other hand some institutions primarily restrict their development activity to the redevelopment of properties in their own portfolio. Involvement in development, whether directly or indirectly, will depend mainly on a particular institution's attitude to risk and their perception of the development cycle at any one point in time. Once again, research into the property market is a critical element that cannot ever be underestimated, especially when the core goal should always be to decrease unnecessary exposure to risk where possible. It should be noted that undertaking a development is generally a riskier proposition than buying a standing investment, due to various factors. Building costs and land values are comparatively low when compared with the prices being currently sought for prime standing investments. In addition, developing a new property provides the institution with an opportunity to tailor a property in the absence of suitable stock available (now or in the near future) on the market. However, it is important to bear in mind that development only represents a small proportion of all institutional property investment.

Banks and building societies

Banks who participate in the funding of development include both UK and foreign clearing and merchant banks. During the 1980s banks became the dominant supplier of development finance as UK banks competed with foreign banks to fill the gap left by the financial institutions as they reduced their exposure to property. The banks extended their traditional role as short-term financiers and became involved in medium-term loans, usually up to 5 years beyond the completion of a development in order to coincide with the first rent review. They were encouraged by the dramatic growth in capital and rental values of property in the late 1980s property boom. During this time banks were also joined by some of the larger building societies (traditionally residential mortgage providers), although only on smaller schemes. Immediately following the last major property crash in the 1990s most banks concentrated on reducing their debts outstanding to property companies and were understandably cautious before forwarding large sums on a hypothetical property development. Following this crash, many property developments remained un-let or unsold for an extended period, although eventually they were sold at a loss (and declared a bad debt) or there was a delay until the market picked up again. Since this period, the area of property development has once again re-established trust amongst lenders, who also conduct due diligence on a property before advancing funds. In addition, most banks now adopt a 'hands on' approach to understanding the property development industry and are assisted by their own valuation surveyors and research teams. Also, the banks are less likely to lend on purely speculative developments without the provision of a sound business case with comprehensive and industry-supported market data and projections. Furthermore, they place substantial emphasis upon lending on prime low-risk developments that are predominantly pre-let. If the agreed proportion in the development can not be pre-let (or pre-sold), often the development is reconfigured to meet market demand, e.g. modify the quality of development.

It must be noted that the banks are in business to make a direct financial gain from lending money, where lending to property companies has always been viewed as profitable, although subject to market downturns. Bank lending may take the form of 'corporate' lending to a company by means of overdraft facilities or short-term loans. Alternatively, a loan may be made to enable a specific development project to proceed or for a developer to retain a development as an investment. To reduce exposure to risk, the banks will use the development or the investment property and/or the assets of the company as security for loans. Property is attractive as security for banks as it is a large identifiable asset with a resale value but, importantly, it can not be sold unless it has a clear, unencumbered title of ownership. However, due to the cyclical nature of property markets there will always be periods where property values will decrease. Accordingly, the banks' willingness to lend money to developers is determined by their confidence in the property market and the underlying economy at any particular time.

The clearing banks, due to their large deposit base, are the major providers of corporate finance loans to development and property companies. Some project finance is provided by the clearing banks on small schemes being developed by established customers. The merchant banks, with some being subsidiaries of clearing banks, and specialist property lenders have smaller funds but have more property expertise. Accordingly, they are more inclined to provide project loans and because of their expertise will take on high-risk loans in return for an equity stake in a project. Merchant banks on large development projects may assume the role of the 'lead' bank and assemble a syndicate of banks to provide finance and in doing so may underwrite the loan. Merchant banks also undertake investment management on behalf of institutional investors through investment funds and unit trusts. In practice the respective roles of both the clearing banks and merchant banks may overlap, particularly if they are associated. Many foreign banks are represented in major cities throughout the world, which also operate in a similar way to British clearing banks.

There are also financial intermediaries who act as agents or financial advisers structuring development finance with banks and other sources for a developer in return for a fee, typically 1 per cent of the value of the loan sought. Several of the large surveyor/agency firms such as CBRE, Jones Lang LaSalle, and Chesterton offer different services to clients, as well as having financial service arms.

Since the 1986 Building Societies Act, building societies have been allowed to provide corporate loans and loans secured on commercial rented property, provided such loans comprise a relatively small proportion of the total loan. In the past some building societies have been exposed to risk due to the ups and downs in property cycles, particularly with residential developers, and as such as have withdrawn from development funding, although other societies have still been willing to fund commercial property investments. Typically they restrict themselves to smaller loans and tend to be less competitive than banks in relation to the interest rates they charge, although this varies between different societies and banks.

When lending money the lender's main task is to assess their exposure to risk, based primarily on the possibility that the borrower/s will not be able to meet agreed regular payments, will not be able to pay back any money, and/or what the minimum amount is that the property could be sold for. The banks' criteria for making loans will vary depending on many unique factors including the nature and size of the development company, the nature and size of the development, the length of the loan and the strength of the security being offered. In assessing the risk of corporate loans the bank will be concerned with the financial strength, property assets, track record, profits and cash flow of the development company. In relation to loans on specific developments the banks will also be concerned with the security of the development project. The banks will seek to assure themselves that the property is well located, that the developer has the ability to complete the project and that the overall scheme is viable. The in-house team of property experts with external advisers, if necessary, will carry out an assessment and valuation of the project. In the case of medium-term loans, where the developer wishes to retain the completed property until the first rent review, and loans on investment properties, the banks will also be concerned as to whether the rental income covers the interest payments. For example, in the late 1980s the banks were prepared to provide loans where the rental income did not cover the interest payments (referred to as deficit financing) on the basis that both rental and capital values were rising rapidly. Since the inevitable downturn that followed, this policy is no longer in place and the banks are seeking to quantify their exposure to risk in both a buoyant and a depressed market.

Generally speaking, banks have always tended to take a short-term view in relation to their lending policies, being concerned primarily with the underlying value of the development company and/or development project. As a result there was widespread criticism of the banks' lending policies during the late 1980s, with many partly laying the blame for the boom/slump at their door. One argument was that the banks relied too heavily on the security and did not pay enough attention to assessing the risk attached to both borrower and projects. On the other hand the banks were quick to blame the valuers whose opinions formed the basis for assessing loans, evidenced by the many negligence cases pending against valuers. Nevertheless, both the banks and their advisers have learnt from their experience of the dramatic fall in property values in the early 1990s. This has led banks to review their policies and take a more medium-term view, beyond prevailing market conditions, accompanied by a higher reliance on research and forecasting methods such as discounted cash flows. Since then the role of the valuation industry has consistently been the focus of debate from a risk perspective, where the valuer is supposedly accountable for any difference in value between the final sale value and the valuation amount on the original valuation date.

Property loans usually account for a relatively small proportion (approximately 10 per cent) of all commercial lending by the banks, and merchant banks are also conscious of their exposure to property. After the last boom/bust cycle in the 1990s many development companies were in default in relation to the conditions of both their corporate and project loans, as the value of their developments/properties were less than their outstanding debts. In order to reduce the losses to their investment, banks were willing to restructure many loans by a combination of measures such as renegotiating loan terms, refinancing loans, swapping 'debt' for 'equity' (converting part of the debt into mortgages) and also forcing sales of assets in worst-case scenarios. In the last market downturn many companies, particularly those with large development programmes and specific projects, went into receivership. Trader–developers were particularly vulnerable and of those who remain many were being tightly controlled by their banks. In a depressed market where the banks decided to stand by and support a developer or project, then the problem of vacant or over-rented property due to the excess supply situation remains. This is further complicated since the institutional investment market is not interested in purchasing over-rented or secondary un-let property. Accordingly, the banks often have to make major write-downs after a downturn and are increasingly reluctant be exposed to this type of risk. More recently the banks have also developed a flexible range of different financing products in order to meet the changing demands of the market and in response to an increasingly competitive financing environment.

Property companies and the stock market

There are two broad categories of company who participate in development: investors and traders. The investor type of company, usually referred to as a property company, is also a source of long-term finance as some purchase property investments for their portfolio as well as retaining their own developments. Their capacity to purchase property depends largely on their ability to raise finance. Property companies and development companies alike are partly financed by their own capital and that of their shareholders, and partly financed by borrowing money either short term or long term. The level of 'gearing' (relationship between borrowed money and the company's own money) will vary between companies. Property companies, as opposed to 'trader' development companies, tend to have a lower level of gearing due to the strength of their asset base.

Property companies vary from small, private firms to large publicly quoted companies. Some specialise in a particular geographical location such as a quadrant in a city, while others hold large portfolios of a cross section of property types in both the UK and overseas. Their prime objective is to make a direct profit from their investment and development activities, while some will take a longer-term view than others to the extent to which they 'trade on' their investments and completed developments. Most importantly, they always have a responsibility to their shareholders to 'increase shareholders' wealth' by maintaining the share price and providing dividends. Property companies view property investment as both a source of income and as an asset providing security for borrowed money. Property companies, particularly the large quoted ones, will tend to concentrate their investment activities on prime and good secondary properties. However, unlike the institutions, they see the management aspects of property investment as an advantage. They have both the management and development skills in-house to improve the value of properties. Also they are not adverse to multi-let properties provided they are in a good location and of high quality. They may purchase investment properties with redevelopment potential that are not fully let or are nearing the end of the lease.

The shareholders of property companies are a combination of financial institutions and private individuals. Financial institutions invest in property company shares instead of or in addition to their direct property investments. However, it is not tax efficient for pension funds (who are non-tax payers) to invest in property company shares when compared with investing in property directly. This is because corporation tax is paid on the company's profits before dividends on the shares are paid and capital gains tax is paid on property sales. Tax-paying shareholders will be taxed on the dividends and on any capital gains from selling the shares.

With a quoted property investment company, the shares are valued by the stock market below the value of the assets of the company attributable to the shares; this is known as the NAV or net asset value per share. This discount to asset value is due to the tax disadvantage of the company as capital gains tax might be payable on the sale of their assets. Importantly, the amount to which shares are discounted varies with stock market conditions and the state of the property market. However, the value of property company shares fluctuates more widely than the value of property, regardless of the state of the property market. This was shown to be the case on 'Black Monday' in October 1987, when the property market was booming but the shares of property companies fell dramatically along with all other equities, although when viewed over the long term there was a correlation between the trends in property values and the share prices of property companies (Rodney and Rydin, 1989). Other factors affecting the value of the share price of property companies are the financial strength of the company including its level of gearing as evidenced by their balance sheet, as well as the perceived strength of the management team. In a similar manner to other companies listed on the stock market, the price earnings ratio (P/E) is the main yardstick used to assess market perception of the future earning potential of property trading companies.

Equity finance can be raised by issuing various forms of shares in a company, with investors directly participating in the profits and risks of the company. New property companies may float on the stock market and raise money by selling shares, often referred to as an IPO (initial purchase offering). Quoted companies can issue new ordinary shares or preference shares to raise equity finance for their development activities, depending on stock market conditions, the overall performance of property company shares and the NAV per share of a particular company. Such finance may also be used to repay bank borrowings and other debts, or alternatively retain strategic developments in an investment portfolio. In addition, companies can also raise debt finance via various methods on the stock market. Long-term debt finance is capital borrowed from investors that usually involves fixed interest, and may be secured on the company or unsecured. Debt finance usually has to be repaid by a certain date or converted into shares (i.e.

equity). Debt finance instruments became popular as an alternative means of providing long-term finance to hold developments as investments.

Real Estate Investment Trusts (REITs)

Real Estate Investment Trusts, commonly referred to as REITS, have been a successful vehicle for the securitisation of property or real estate in many countries including the US, Australia and Singapore. The increased popularity of REITs is linked to many advantages including taxation incentives, availability of up-to-date information about the REIT and being trading on the central stock market. It has been advocated that REITs help to facilitate the development of a high-quality residential letting market and offer the prospect of boosting house building over the medium to long term (Royal Institution of Chartered Surveyors, 2007a).

REITs were introduced into the UK on 1 January 2007 in accordance with the Finance Act 2006. The UK REITs have many of the benefits of other REITs including greater flexibility and liquidity. However, one of the most sought after benefits was from a taxation perspective – for example, UK REITs are treated as normal corporate vehicles that make an election to confer exemption on taxation from relevant company profits, and in return the REIT must withhold tax from distributions paid to shareholders out of these profits (KPMG, 2007). The requirements to qualify for a UK REIT are as follows:

- UK tax resident (and not dual resident);
- listed on a recognised stock exchange;
- not an open-ended investment company;
- the only classes of shares allowed are ordinary shares (one class only) and non-participating preference shares;
- distributes 90 per cent of its net taxable rental profits (not capital gains) during the relevant accounting period or within twelve months of its end;
- derives at least 75 per cent of its total profits from its tax-exempt property letting business;
- at least 75 per cent of the total value of assets held by the REIT must be held for the tax-exempt property letting business;
- additional conditions also apply (KPMG, 2007).

Whilst a REIT has been a widely accepted vehicle for funding a property via listing on the stock market, there are limitations that should be acknowledged. For example, expenses associated with listing on the stock market are substantial and include the marketing and statutory charges, in addition to the risk that the IPO will not be fully subscribed by investors. Also, investment in direct real estate in a buoyant market may offer a higher yield at times (Reed, 2007), therefore a REIT could struggle to offer a competitive yield regardless of tax advantages (Royal Institution of Chartered Surveyors, 2007b). Over time many property developers have grown from relative small developments and are now listed as a global REIT (e.g. Multiplex, Westfield), although the economies of scale are reserved for larger property developers.

Overseas investors

Due to the widespread effect of globalisation and assisted by advances in information transfer including the internet, overseas investors have become significant participants in the property investment market. No longer is demand for property limited to a prospective purchaser's geographic location – it is commonplace for an overseas investor to be just as fluent with a property market across the other side of the world as with their own local market. At times there are other reasons, such as tax implications or a lack of perceived local market demand, as to why an investor may be interested in a market in an overseas country. For example, in the past overseas investors were attracted to the UK property market due to the unique existence of the 'institutional' lease offering security of income for up to 25 years.

The source of overseas investment has varied since 1988. The Japanese and the Scandinavians led the way during the boom period between 1987 and 1990. The UK government's favourable treatment of foreign investment together with the lifting of the Japanese government's restrictions provided the impetus for Japanese investors, developers and contractors. The lifting of restrictions by the Swedish government on overseas investment by their property companies and life funds led to the Scandinavian interest. They were both attracted by the performance of the UK's economy and its relative stability at that time. Their development companies became involved in direct development either in partnership or on their own account. However, many have since gone into receivership following the collapse of the market in 1990, which followed the high-profile Asian economic crisis.

European investors, particularly the Germans, have become significant since 1992 due to the relative performance of the UK economy against their own economies and the continuing deregulation of cross-border investment by the European Union. They were joined by American, Middle Eastern and Far Eastern investors in the years of 1993 and 1994. All have tended, in contrast to the Japanese and the Scandinavians earlier, to invest in standing investments rather than developments. Like the UK institutions, they are interested in 'prime' properties let on institutional leases to good covenants.

More recently global investors have not been restricted from any one particular country and this trend in many ways mirrors the explosion of information transfer and availability, primarily due to the internet. In a relatively short period of time internet marketing has become a prime advertising medium, which in turn has allowed an investor on the other side of the world to access detailed information about proposed and existing real estate developments. Importantly, this includes digital photographs, three dimensional videos and virtual demonstrations that assist overseas investors to commit substantial funds, even though they possibly have never laid eyes on the property in person.

Expanding the market to overseas investors also has other benefits. This includes increasing the number of prospective purchasers by enlarging the marketplace, increasing the potential borrowing capacity of the purchasers (as opposed to the lenders in the local market) and ensuring the development is perceived as truly international. Furthermore, in the UK many overseas investors are more likely to invest in larger properties at the higher end of the market, which is often the upper limit of all but a few of the UK institutions.

Private individuals

The majority of private individuals purchase property investments at the lower end of the market, with a large proportion being 'mum' and 'dad' investors who are borrowing against the equity in their principal place of residence in the form of the family home. In many instances they tend to concentrate their purchases on secondary and tertiary commercial property with high yields but often located in an area with the perceived potential for capital growth. They tend to be precluded from the 'prime' market due to the significant sums of money involved and the existence of the 'reverse yield' gap. Private individuals are attracted to high-yielding properties as income will very often be in excess of interest rates. However, participation in the lower end of the property market is very risky, involving intensive management and regular voids.

Unfortunately, many private investors will often focus too much on the relationship between return and capital outlay, and make a direct but false comparison with the return from a standard bank deposit. This is partly due to the lack of understanding about the fundamentals of property investment, including the depreciation of the building component over time and associated risk. For example, many investors are not fully conversant with the risk reflected in the yield rate, where a higher yield equates to a higher risk, not lower. In such instances, with a narrow focus on the yield only (and ignoring the long-term maintenance and upkeep costs), some private investors have a smaller initial outlay but a substantially larger cost of maintenance. This is another reason why smaller investors are predominantly at the bottom end of the perceived 'bargain' market for investment properties.

Joint venture partners

A development company may raise finance or secure the acquisition of land by forming a partnership or a joint venture company with a third party to carry out a specific development or a whole series of development projects. The basic principle behind forming a partnership from the developer's point of view is to secure either finance or land, in return for a share in the profits of the development scheme or the joint venture company. The joint share given to the third party will depend largely on the value of their contribution combined with the extent they wish to participate in the risk of the scheme.

There are many forms and methods of forming partnerships or joint ventures for the purpose of successfully funding and completing a property development. Nevertheless, it is beyond the scope of this book, as an introductory text, to examine them in detail and it is recommended that the reader consult a specialised text. However, we will briefly examine the reasons behind forming partnerships and joint ventures, with some examples.

A partnership may involve any combination of sharing the risks and rewards of a scheme via many different contractual and company arrangements. In addition, the partners to a scheme may take an active or passive role in the scheme. Tax and financial considerations may determine to a large extent the formal structure of the partnership arrangement. A joint venture may also take many forms, but in its 'purest' form the parties participate in the development and distribute the profits in equal shares usually by forming a joint company.

Developers are typically reluctant to share profits with third parties, unless it is the only way of securing a particular site or finance for a development scheme. Partnerships with landowners may be required if the landowner wishes to participate in the profits of the development scheme or wishes to retain a long-term legal interest in the property, preferring income to a one-off capital receipt. Local authorities and other public bodies, for example Network Rail with reference to railway line infrastructure, may only grant long-term leasehold interests to developers due to their need to retain an underlying continuing interest in the property for financial or operational reasons. It should be noted that there are restrictions placed on local authorities in relation to capital receipts and the forming of joint venture companies. Most often the ground rent and profit-sharing arrangements will be determined by the amount of risk the landowner wishes to bear.

If a developer is involved with a particularly large or complex development, then a prudent way of undertaking such a scheme is to spread the risk through a partnership arrangement. However, this arrangement will often be beyond the financial capability of all but the very largest companies. With such schemes one or more partners may be involved and may include the landowner, contractor or another development company. Developers may also form joint venture companies with other development companies who may have the expertise or experience required for a particular type of development, which is seen as vital to the success of the scheme. There are also examples of developers forming joint venture companies with retailers to combine their respective experience and market knowledge.

During the late 1980s joint venture companies were formed on some of the larger developments to enable the partners to arrange 'limited' or 'non-recourse' finance off balance sheet so the borrowings did not appear on either of the partner's respective balance sheets. The rules were tightened up on these arrangements and the opportunities to benefit from such schemes has reduced substantially or been eliminated.

Regardless of the reasons for forming a partnership to finance a scheme, it is essential for the developer to ensure the definition of the profit is clearly detailed and understood. In the case of a joint venture company the profit will be distributed through the company accounts. A developer must rehearse every possible outcome of the scheme to ensure that any partnership arrangement will work and the true intentions of the parties have been agreed carefully documented by all involved.

Government assistance

At any given time there are different government grants available, largely dependent on available funding and the perceived ability of government assistance to be a catalyst for change. Many of these projects are designed to encourage private developers to proceed with a proposal, such as renewal or gentrification of an older building or geographical area. In many cases the financial incentives would not be sufficient for the developer to undertake a viable project, although the government acknowledges there would be wider community benefits if a developer would proceed with the project. Examples include the construction of low-cost rental housing or a new shopping centre that would provide local employment. It is beyond the scope of this book to list the individual schemes and incentive programs available, but the developer is strongly encouraged to contact their local government body, which is usually willing to provide information about current and planned future schemes. English Partnerships in the UK has traditionally been an example of the type of organisation that is seeking to assist the community by offering incentives. Its motto is 'supporting high quality sustainable growth in England' and is commonly referred to as the UK's national regeneration agency. Previous and existing schemes include the 'housing gap' funding programme, as well as offering funding for land and property/physical regeneration schemes in different regions (English Partnerships, 2007).

Discussion point

What are the main types of finance available for property development?

4.3 Methods of development finance

A number of different sources for development finance have already been discussed. However, there are many methods of obtaining development finance from the above sources and in recent years there has been the emergence of an increasing number of innovative techniques. It is important to examine the various well-established methods and briefly look at the different finance options available.

The choice of both source and method of development finance will depend on how much equity (the developer's own capital) that the developer is able and willing to commit to a scheme. If the developer has insufficient capital then the aim is to arrange as much external finance as possible in order to meet all costs associated with the property. At the same time the priority is to retain as much of the equity as possible without giving away bank or personal guarantees. A decision has to be made as to how much risk the developer wishes to pass on to the financier in return for a share in the financial success of the scheme. The availability and choice of finance will depend largely on variables such as the company's size, financial strength, track record, characteristics of the development scheme to be funded and the duration of the scheme. Whichever method is chosen, the developer will always need to be fully conversant with all aspects of taxation.

Forward-funding with an institution

Forward-funding is the term given to the method of development finance that involves a pension fund or insurance company agreeing to provide short-term development finance and to purchase the completed property as an investment. This happens at the start or, at least, at an early stage in the development process. This method of finance reduces the developer's exposure to risk, where the terms usually agreed with the institution reflect this. From the institution's point of view, this method of acquiring a property investment has several advantages over purchasing a ready-made investment. From the outset it provides the institution with a slightly higher yield, reflecting the slightly greater risk. By being involved in the development process, the institution influences the design of the scheme and the choice of tenant. In addition, if there is a rise in rents during the development period then the institution benefits from the rental growth.

As previously discussed (see 'Financial institutions'), the proposed development must fall into the 'prime' category if the developer is to be successful in securing forward-funding, although if the proposed scheme is very large in terms of its lot size, then the number of funds in the market is reduced. Therefore, it is essential for developers to fully consider this fact when purchasing a site as it will affect the way in which they evaluate the development opportunity.

On the assumption that the developer is purchasing, or has just purchased, a site in a prime location with the benefit of a planning consent, then the next step is to approach the institutions directly or via their agents. Agents have an important role to play in the forward-funding of a scheme as they have a good knowledge of the institutional investment market. In addition, because so many of them are retained by institutions themselves they have established good contacts over time. The developer may also have established a good working relationship with particular institutions that they may have worked with in the past. Institutions themselves will tend to adopt a proactive approach and directly or through agents seek out the development opportunities themselves in accordance with their individual investment criteria. They will have identified through their research the property type and location they are interested in.

The developer will usually prepare a full colour presentation brochure for those institutions who express initial interest in the scheme. The brochure will typically describe the nature and location of the development, with supporting illustrative material – in addition it outlines planning consents, site investigation reports and specifications. In this process the institutions will seek to identify the level of risk associated with the potential development, including the appeal to purchasers and the underlying level of demand that the completed development could generate.

Most importantly, developers will need to sell their track record and experience on similar schemes. For example, there will be an additional risk component attached to a relatively new developer who does not have the perceived ability through their development history to undertake and successfully complete a property development from start to finish. It is essential that the developer provides an analysis of the market in terms of the balance between supply and demand (see Chapter 7) for similar schemes. An initial appraisal of the scheme will also be included as a starting point for negotiations on the value and cost of the proposed scheme, with any supporting evidence such as cost plans. However, the institution will not rely on either the market analysis or appraisal by the developer and will carry out their own evaluation to assess the risk, using a variety of sensitivity techniques (see Chapter 3).

The institution, having decided the proposition fits within its investment criteria, will need to satisfy itself that the proposed development is viable, that there is a demand for the development, that the specification and design of the building is of the highest quality, and that the developer has a satisfactory track record and expertise. Once again the institution will be constantly reassessing its level of exposure to risk and seeking to reduce it, if possible. The developer may need to be able to guarantee the investor's return at the end of the development period depending on the terms negotiated, so the institution will need to examine the developer's financial standing. The institution will ask itself the following question: 'Is the developer able to produce the scheme both within budget and on time?'

After a particular fund has agreed in principle to the forward-funding of the developer's scheme, negotiations can then commence about the financial aspects of the agreement. There are various types of arrangements that can be entered into largely depending on variables including the current state of the property market, the nature of the scheme and the financial standing of the developer. Funds will tend to tailor the arrangements to suit the particular development and their view of the market. There are many characteristics that are typical of most deals and the variation between arrangements will be reflected in the balance of risk and reward between the parties.

We shall now examine each of the typical elements of a funding agreement.

Yield

The fund and the developer will agree at the outset the appropriate yield (and therefore capitalisation rate) applicable to the scheme. This is commonly referred to as the relationship between the total capital value and the net operating income. The yield will usually be determined after a thorough assessment of recent market evidence and the fund's perception of the current and future risk. In valuation circles, the yield or capitalisation rate is also referred to as the 'all risk yield' since it is supposed to reflect all risk associated with the property. As previously discussed in Chapter 3, the yield is a measure of the institution's perception of risk weighed against the rental growth prospects. In forward-funding arrangements yields will normally be discounted by around 1–2 per cent (i.e. 1–2 per cent higher) from the market yield for standing investments to reflect the additional risk the institution is taking by participating in the development process. The yield will be fixed at the agreed level. It is critical that the property developer has a clear understanding of how yields are calculated and, importantly, the relationship between the yield rate and risk.

Rent

What is commonly referred to as a 'base rent' will be agreed upon after a careful analysis of current market evidence, a comparison of the newly constructed floor space to recent lettings of similar properties in the area. At times there will be a pre-letting in place with the rental level agreed with the tenant in the 'agreement to lease', although there may be provisions for a review on final completion of the property development; this would be more likely to occur when there is an extended time period, e.g. a large-scale development. If the rent achieved on the scheme exceeds the base rate then there is normally a provision to share this benefit. This element of the rental income is known as the 'overage' and is usually shared between the fund and the developer. However, the fund may cap the 'overage' rent at a certain level because the developer will be motivated to achieve the highest rent and the fund will wish to safeguard against 'over-renting' the property, which would be detrimental to rental growth prospects in the future. In other words, there is a trade-off between achieving the highest possible level of rent and retaining a long-term tenant. Usually such a tenant will quickly relocate to less expensive accommodation at the first available opportunity, thereby creating an increased void.

Costs

It is essential that the developer presents to the fund a detailed estimate of current and reliable development costs. These will be analysed by the fund's in-house building surveyors or externally appointed consultants. Possibly the institution will wish to cap the total development costs at a certain level, although allowing for interest. The developer will be under an obligation not to exceed these costs. If the maximum agreed limit is exceeded then the developer will be responsible for funding the balance. There may be a provision within the agreed development cost for the developer's own internal costs, such as project management fees, overheads and a contingency allowance. The maximum agreed development cost will typically relate to previously agreed plans and specifications. Note that the developer will be under an obligation not to vary either the plans or specifications without the prior approval of the fund. There may be a provision that enables a variation of the agreed costs due to agreed variations.

The fund will provide the short-term finance at an interest rate to reflect their opportunity cost of money and not the cost of borrowing, so in practice the rate has tended to average 5–8 per cent. This is because institutions do not need to borrow money, but regard the provision of short-term development finance as part of their investment. Money will be advanced to the developer on a progress basis, commencing with the production of architect's certificates in respect of the building costs and invoices in respect of all other costs. Interest will accrue and be rolled up until practical completion or until the scheme is fully let depending on whether the developer is responsible for any shortfall in rent. At the same time the developer will receive any profit, calculated as the development value less the development costs advanced in accordance with the terms of the funding agreement, although the fund will keep a retention that is equivalent to or greater than the amount agreed with the building contractor, depending on the existence of any defects or work outstanding.

Depending on market conditions, the developer may be able to secure a profit on the value of the land if the value of the land at the time of the funding agreement is greater than the initial cost of acquisition.

Developer's profit

The calculation of the developer's profit can vary and ultimately will depend on the type of funding arrangement entered into. The developer and fund will agree either a base rent or a priority yield method of funding.

1. **'Base rent' arrangement**   On completion of a scheme based on a base rent arrangement the total development cost (including interest) up to any maximum agreed limit will be deducted from the agreed net development value for the scheme. The net development value is the total rent achieved up to the agreed base rent multiplied by the agreed year's purchase (a reciprocal of the yield) less the institution's costs of purchase. The balance of the calculation will represent the developer's profit, often referred to as a 'balancing payment'.

As an example, we will look at the evaluation in Chapter 3 on the assumption of a forward-funded deal with an agreed base rent. Assume that the developer has agreed with the institution a base rental value of £1,032,573 per annum, based on a net lettable area of 3553 m^2 (38,250 ft^2) at a base rent of £290.62 per m^2 (£27 per ft^2) and a yield of 7.5 per cent. It is also agreed that any rent achieved above the base rent will be split evenly with 50 per cent to the developer and 50 per cent to the institution. If the rent achieved is £1,200,000 per annum and the development cost is £12,000,000 then the developer's profit is calculated in the way shown in Example 4.1.

The fund's profit is represented by the movement in the initial yield from 7.5 to 8.29 per cent calculated by dividing the rent achieved by the development value and multiplying by 100, i.e. (£1,200,000/£14,470,895 × 100 = 8.29 per cent).

Example 4.1   Developer's profit calculation (1)

Base rent plus (per annum)	£1,032,573 p.a
50% overage (i.e. £1,200,000 p.a. - £1,032,573 p.a. divided by 2)	£83,714
	£1,116,286
Capitalised at 7.5%	13.33
Gross Development Value	£14,880,098
Less purchaser's costs @2.75% equals	£14,470,895
Less development cost	£12,000,000
Balancing payment to developer (i.e. developer's profit)	£2,470,895
Developer's profit as a percentage of cost	20.59%

---

2. **'Priority yield' arrangement**   A 'priority yield' arrangement provides the fund with the first or 'priority' slice of the rental income before the developer takes a profit, providing a guaranteed return yield on the institution's investment. The priority yield is used particularly where the costs or rents are perceived to be subject to a greater uncertainty (e.g. with lengthy schemes) and, therefore, present a greater risk.

Alternatively, the fund will agree with the developer the priority yield – in Example 4.2, 7.5 per cent – and will receive as a priority slice 7.5 per cent of the development costs. Then the developer will receive as the next slice an agreed percentage of the development cost – in Example 4.2, 1 per cent – which is then capitalised at the agreed base yield. This equates to the developer's required profit as agreed with the fund. The remaining rental income is split 50/50 or as otherwise agreed, then capitalized at the agreed base yield. Again using the evaluation in Chapter 3 we now rework Example 4.1 on the basis of a priority yield arrangement. The developer's profit is calculated in the way shown in Example 4.2.

(The fund's profit is represented by the movement in the initial yield from 7.5 to 8.29 per cent as in Example 4.1).

Example 4.2   Developer's profit calculation (2)

Development cost	£12,000,000
Achieved rental income (per annum)	£1,200,000
Fund receives first slice of rental income	
@ 6.5% of development cost	£780,000
Developer receives next slice of rental income	
@ 1 % of development cost	£120,000
Developer/fund share balance of rental income 50%-50%	
(i.e. £83,714 divided by 2)	£41,857
Developer's profit is the share of rental achieved	£161,857
Capitalised at 7.5%	13.33
Developer's profit	£2,157,551
Developer's profit as % of cost	17.98%

Developer's guarantees and performance obligations

In addition to controlling their maximum funding commitment by capping costs, the fund may require certain guarantees from the developer. Unless the scheme is entirely pre-let then the fund may require the developer to guarantee any shortfall in rental income until the scheme is fully let or 3 to 5 years after completion, whichever occurs first. Alternatively, the fund may require the developer to enter into a short-term lease for 3 to 5 years after completion. Bank guarantees or parent company guarantees may also be required to support the potential rent liability to meet any costs exceeding the agreed limit.

If the developer wishes not to provide such guarantees or enter into a lease then a 'profit erosion' arrangement may be entered into. Under this arrangement interest and costs will continue to accrue until the scheme is income producing or 3 years after completion, whichever is the earlier. At this point any profit due to the developer will be calculated and may have been entirely eroded through an increase in the development costs above any agreed limit or a decrease in the rent actually achieved. However, the developer will then be able to completely walk away from the development without any further commitments.

Typical of most funding arrangements is an obligation by the developer to perform – to build the scheme in accordance with the agreed specification and plans within the time and budget agreed. In addition, the developer has to ensure that the professional team performs and that collateral warranties are procured for the benefit of the fund (see Chapter 6 for further details on collateral warranties). Throughout the development the fund's interest will be protected by their surveyor who will oversee the project, attending site meetings as an observer, to ensure the developer is performing in accordance with the agreed specification and plans.

Lettings

The developer will need to obtain the approval of the fund for all lettings. The funding agreement will usually specify on what terms the fund will be prepared to grant leases – a standard form is usually attached to the funding agreement. There may be a provision that allows a particular type of lease – for example, over a particular period of time with agreed rent reviews. However, increasingly more flexible lease lengths are allowed with or without breaks – from a lessor's perspective the longer and more rigid the lease, the less risk of a void. On the other hand, from a lessee's perspective it is important the lease is as flexible as possible to allow for changing circumstances (e.g. more staff) in the future, which would decrease the lessee's risk. The agreement may also specify an acceptable tenant. A typical arrangement may specify that the tenant's profits for the last 3 years must exceed a sum three times the rent or total liability (including service charges). The fund may specify that only single lettings are acceptable, although floor-by-floor lettings might be acceptable after a certain length of time from practical completion.

Sale and leaseback

As an alternative to the above described arrangements, a sale and leaseback arrangement may be entered into whereby the developer retains an interest in the investment created by the development. This type of arrangement varies depending on variables such as the availability of land, the amount of money available and so forth. However, the arrangement is not suitable for all stakeholders, for example where institutions prefer to retain total control and choose the leasing arrangements. A sale and leaseback involves the freehold of the scheme passing to the fund on completion, with the fund simultaneously granting a long lease to the developer, who in turn grants a sublease to an occupational tenant. There are many variations of this arrangement depending on the method of sharing the rental income. Sale and leaseback arrangements may be either 'top sliced' or 'vertically sliced'. With the top slice arrangement the fund receives rent from the developer in accordance with the required yield. The developer is then able to retain any profit from letting the property at a higher rent than that payable to the fund. However, with upward-only rent reviews in the developer's lease with the fund the developer's profit rent may be rapidly eroded over time. This means the developer's interest is only saleable to the fund. Therefore, the 'vertical slice' arrangement is better from the developer's viewpoint as the fund and the developer share the rental income from the property in relation to an initially agreed percentage throughout the length of the lease.

Institutions are more likely to enter into sale and leaseback arrangements directly with the occupiers to create attractive property investments with tenants of good financial standing.

When undertaking a property development, the cyclical nature of the market is a major consideration and the state of the market at any one particular point in time must be considered. In a property market where there is an oversupply of space (mainly secondary) many institutions would then enter into forward-funding deals on the basis of a pre-letting. Alternatively, they would be prepared to enter into arrangements with a developer of good financial standing where the rent is guaranteed for 3 to 5 years by the developer. However, forward-funding deals are currently being achieved on the basis of speculative schemes, providing there is a lack of 'prime' space on the market and there is proven demand, where the risk is quantifiable by the institution.

Bank loans

In recent years the banks have had increased competition due to the globalisation of the banking sector, as well as other types of lenders entering the market with hybrid products. The clearing and merchant banks provide short-term development finance, either on a 'rolling' or project-by-project basis, by means of overdraft facilities or short-term corporate loans secured against the assets of the development company or project loans secured against a particular development. With the dramatic increase in bank lending in recent years and the wider acceptance of debt, various different methods of bank lending have been introduced, such as development companies seeking bank finance beyond the construction period up to the first rent review. Another example is the popularity of mezzanine finance, which is increasingly filling the gap between equity and a first mortgage.

For many developers, especially the smaller ones, forward-funding is difficult to obtain as they are unable to provide the requested guarantees. Also 'prime' properties that are acceptable to institutions represent a very small part of the market and to some extent tend to be geographically restricted to London and towns in the South East. Large development projects with extended planning and construction phases are beyond the capacity of all but a few of the larger funds. For instance, if a development company is seeking to forward-fund a retail town centre scheme worth in excess of £250 million, then they may need to involve multiple funds, and the choice is very limited. From the developer's point of view, borrowing from a bank allows greater flexibility and enables the developer to benefit from all of the growth, unless some of the equity is given away. The developer can repay or refinance the debt when the time is right and sell on the completed investment at a higher price. In addition, the developer will not be subject to the same degree of supervision through the development process. In a rising market where rents and capital values rise rapidly, it is more profitable for developers to arrange debt finance as opposed to equity finance for the reasons mentioned above.

Corporate loans

Development companies can arrange overdraft facilities or loan facilities with clearing banks secured on their assets. However, with corporate lending the bank is concerned with the strength of the company, its assets, profits and cash flow. Accordingly, obtaining bank loans in this way is more appropriate to investor–developers and large developers rather than trader–developers and smaller developers as they have large asset bases that can provide the necessary security for bank borrowings. Usually corporate loans can be obtained at lower interest rates than project loans.

Project loans

Alternatively, development companies can arrange project loans that are secured against a specific development project. Banks normally provide loans that represent 65–70 per cent of the development value or 70–80 per cent of the development cost. Developers have to provide the balance of moneys required from their own resources. The banks limit their total loan to allow for the risk of a reduction in value of the scheme during the period of the loan. In addition, by insisting on an equity injection by the developer they are committing the developer. This equity provision is normally required at the outset of the development to motivate the developer to complete the scheme. Also the developer is totally responsible for any cost overruns. The loan to value ratio (LVR) depends on the risk perceived by the bank and can vary substantially depending on the risk profile of the borrower, the perceived risk in the project and prevailing market at the time. Clearly a pre-let or pre-sold development represents less risk than a totally speculative one. At times it has been possible to secure between 85 and 100 per cent of development costs through various layers of bank finance using a combination of insurance, mezzanine finance and profit-sharing arrangements. However, in periods of sustained downturn this high LVR would be considered too risky by the financier. Lending conditions will vary according to the banking sector's knowledge and overall confidence in the property market (systematic risk) and in the actual property itself (unsystematic risk).

Project loans are attractive to the smaller trading companies as the companies are not worth enough to fund their full development programme through corporate loans. For many larger companies, these loans can be carried off the parent company's balance sheet by forming joint ventures with the bank in a subsidiary company; the borrowing associated with the property would not appear on the parent company's balance sheet. This enabled development companies in the 1980s to have development programmes that would have otherwise been impossible as gearing would have increased to an unacceptable level. In this example, gearing is commonly referred to as the relationship between borrowed money (liabilities) and the company's own money (equity). At times some investments are negatively geared, such as where outgoings exceed income – note that in some countries this loss can be offset against other income to reduce tax.

When the banks lend against a particular development project, it will form all or part of the security for the loan. The developer needs to provide similar information to the bank as to a funding institution. The banks will wish to ensure that the property is well located, that the developer has the ability to complete the project and that the scheme is viable. This requires the bank to have knowledge of the property market either through in-house staff or external advice from firms of chartered surveyors in order to assess the risk involved. The developer will need to present the proposal to the bank in the form of a package very similar to that required by a funding institution (see 'Forward-funding with an institution').

However, as the bank is viewing the scheme as a form of security and not investment, it is more concerned with the underlying value of the scheme rather than the details of the specification. The appraisal, therefore, forms the most important part of the presentation together with all the supporting information and market analysis. It must reflect all of the risk in the proposal with supporting market research to reduce the unknowns. Equally important is the track record of the developer in carrying out similar schemes. The banks will also examine the financial strength of the development company, although this may not form part of the security of the loan.

The bank will employ either its in-house team of experts or external surveyors/valuers to report on the proposal and provide a valuation of the proposed scheme. Part of the process will be an analysis of the risks involved and this should be reflected in the terms offered to the developer. The bank will be concerned to ensure there is sufficient contingency and profit/risk allocation built into the appraisal to provide a sufficient margin for cost increases during the period of the loan.

Another element of the risk involved in bank loans is represented by fluctuations in interest rates, which the bank may be concerned to limit (see below). Interest rates on bank loans can be at a fixed percentage, a variable percentage or a combination of both – if the rate is fixed it is only in relation to the base rate or LIBOR (the London Interbank Offered Rate, i.e. the rate of interest between banks). This proportion is agreed between the bank and the developer, in other words that the interest rate is, say, 1.5 per cent above LIBOR. Generally interest rates will be higher on project finance loans than corporate loans due to the higher uncertainly and, therefore, the increased risk that the banks face. The interest rate margin may be less if the developer pre-lets or pre-sells the property before completion as this substantially reduces the bank's risk. Also the interest-rate margin on an investment loan would be lower than on a speculative development loan. Interest rates on short-term loans are likely to be floating whereas on long-term loans they are more likely to be fixed.

The bank should also consider how the loan will be repaid either by the sale of the completed scheme or by refinancing from another bank. It must be remembered that on completion of the scheme the initial rental income will usually be insufficient to cover the interest costs on the loan, due to the 'reverse yield' gap problem where yields on property investment tend to be lower than medium to long-term interest rates.

The bank will also need to protect its 'security' by obtaining a first legal charge on the site and development – it is important for this to be recorded on the title or deed to the property and, therefore, any prospective purchasers undertaking a search will be aware of this liability before buying the property. In the event of default on the loan the bank will be able to obtain ownership of the development if required. It may also require a floating charge over the assets of the development company. The bank will need to be able to be legally capable of stepping into the 'developer's shoes' in the event of any default that may require the legal assignment of the building contract and any pre-sale or pre-letting agreements. Like a funding institution, the bank will also require collateral warranties from the professional team.

Guarantees may be required from the parent company or a third party if the financial strength of the development company is not considered adequate due to the perceived higher risk. A full recourse loan will involve the parent company providing a full guarantee on the developer's capital and interest payments together with a guarantee that the project will be completed.

Previously limited and non-recourse loans became an attractive proposition for developers wishing to finance their development projects whilst providing limited or no guarantees. With a 'limited recourse' loan the parent company may only have to guarantee cost and interest overruns. Limited recourse loans were normally granted for the construction period of a project and up until the first rent review. A 'non-recourse' loan involves no guarantee with the only security for the bank being the development project itself. However, in practice the parent company is still responsible and it would be very difficult for a developer to simply walk away from the scheme without damaging their long-term reputation.

The developer needs to take account of the considerable costs involved in bank finance, which are usually upfront. The developer will need to pay for the cost of carrying out their own appraisal and presentation to the bank. In addition, there are several fees payable to the bank, although some fees may be negotiable depending on the size and type of the loan. For example, if multiple banks are competing to fund a large development project, one bank may lower or remove the fees altogether to increase their competitiveness. In addition, an arrangement fee will normally be charged by the bank to cover the cost of carrying out a valuation and assessment of the project. This fee may include an element of profit depending on the risks involved. There will also be a management fee to cover the bank's costs in monitoring the project, consisting of mainly surveyor's fees. Such charges can represent 3–10 per cent of the value of the loan. In some instances there may be a 'non-utilisation' or 'commitment' fee on the part of the loan not drawn down initially as the bank will have to retain the full loan facility and cannot commit the funds elsewhere.

There are some variations on the basic project loan described above.

1. **Investment loans**   Development companies, wishing to retain a development, can secure the option to convert the project loan into an investment
   loan on the completion of the project once it is fully let, usually up until

the first rent review. Alternatively, the developer may agree a combined project and investment loan from the outset. Also the developer may be able to refinance a loan on completion on better terms than a previous short-term development loan. On investment loans, banks will normally lend up to three quarters of the value but there is always the problem of the rent not covering the interest payments. Banks wish to see the interest covered by the rental income and, therefore, can limit the loan. This may be relaxed where the property is reversionary (let below market value) or the parent company guarantees the shortfall interest. Otherwise the banks may require the developer to cap the interest rate or rearrange the payments on the loan (see below). The interest rates on investment loans are usually lower than on substantially riskier speculative project loans and the risk to the bank will depend on the financial standing of the tenant.

2. **Mezzanine finance**   A project loan may be split into different layers known as 'senior' debt and 'mezzanine' debt if the developer is unable to meet the normal equity requirement or wishes to increase the amount of the loan above the normal loan-to-cost ratios. The senior debt usually represents the first 70 per cent of the cost of the development scheme, like a straightforward project loan. Senior debt is usually provided by the major UK clearing banks and is commonly referred to as the first mortgage, since it takes priority over other forms of debt if the property is sold to reclaim funds. This debt may represent more than 70 per cent if the development is pre-let or pre-sold. When a developer wants to borrow more of the cost of the project than 70 per cent, additional money may be raised in the form of mezzanine finance. Also the bank may increase the senior debt exposure to 85 per cent of cost with a commercial mortgage indemnity scheme. Mezzanine finance may involve the developer losing some of the equity or, alternatively, taking out an insurance policy. This mezzanine element is normally provided by merchant banks and specialist property lenders. As this mezzanine level of finance is more risky and often referred to as a second mortgage, the bank will charge a higher interest rate (usually 1–2 per cent higher than the rate on the senior debt) or require a share in the profits of the development. The banks will also require full guarantees from the parent company.

If a mortgage indemnity insurance policy is taken out it will reimburse the lender if the loan is not repaid in full and will involve the developer paying a substantial one-off premium to a specialist insurance company. The policy may cover the mezzanine layer of the loan or the entire loan. Equity sharing with the bank may involve a profit share or an option on a legal interest in the scheme. Those banks willing to participate in the equity of the scheme are limited to those with sufficient property expertise. Very often, due to the tax complications, the profit share will be expressed as a fee. In this instance, the bank will become part of the development team and become involved in the decision-making process.

3. **Syndicated project loans**   The necessary development finance may need to be borrowed from more than one bank. In particular, larger loans are more likely to be syndicated among financiers or a group of banks by a 'lead' or agent bank. Each bank shares, in proportion, the risk of the development project depending on their initial contribution; their profit is also commensurate with the proportion of their initial contribution. As a further complication, the loan may be just senior debt or it might include a layer of mezzanine finance. The lead bank, usually an established property lending bank with the necessary expertise employed in-house, will arrange the syndication of banks. The lead bank may underwrite the entire facility or agree to use its 'best endeavours' to secure the syndication. It is usual for the lead bank, who may participate in the syndicate, to have the final responsibility for making decisions on behalf of the syndicate during the period of the loan.

4. **Interest rate options**   The development company may wish to protect themselves from the risk of changes in the base rate or LIBOR over the period of the development project, particularly when the market is uncertain or there are indications that in the future interest rates may rise. In this instance, the developer may seek interest loans at a fixed rate, but it must be remembered that the development company may be tied to a high rate of interest for a long period, so that they may be unable to gain from subsequent reductions. On the other hand, a fixed interest rate can remove uncertainty for the lender regardless of external, unknown forces affecting the interest rates. Alternatively, developers may compromise and try to 'hedge' the risk that interest rates will rise during a development, but this will always be at a cost. The usual form of interest rate hedging is the 'cap', which limits the amount of interest the developer will have to pay and is similar to an insurance policy. It is an interest rate 'option' that has to be paid for at the outset, either to the bank providing the loan or to another bank altogether. For instance, with a loan linked to LIBOR, the bank will reimburse the developer the cost of interest over and above the cap rate. Hedging is more difficult on a speculative development loan than on loans for income-producing investment properties due to the uncertainty of the amount of loan outstanding at any one time.

Mortgages

Mortgages provided the most common form of long-term development finance until the 1960s. A mortgage is a loan secured on a property whereby the borrower has to repay the capital loan plus interest by a certain date. However, not many lenders are interested in long-term non-equity participating loans such as mortgages. From the lender's point of view, a mortgage is a fixed-income investment and very illiquid. Some banks, the larger building societies and many life or insurance companies provide mortgages on commercial properties. However, the availability of mortgages is limited due to the problem of the reverse yield gap. It is rare to get fixed-rate-interest mortgages, although some life funds provide long-term, fixed-rate-interest mortgages depending on prevailing interest rates. Demand for mortgages has been mixed over the past due to changing economic circumstances – for example, mortgages regained some popularity in the late 1980s as long-term interest rates were low, but by the early 1990s long-term rates rose again, although since the mid-1990s interest rates have remained relatively low again.

Mortgages may normally be granted on a loan-to-value basis of between 60 and 80 per cent depending on the risk involved. The amount of mortgage secured will depend on the security being offered by the borrower in relation to the quality of property, the financial standing of the tenant and the borrower. Mortgage loans are normally 20–25 years in line with the length of occupational leases, although this time period is open to negotiation.

Various methods have been developed to overcome the initial deficit problem caused by the difference between rental income and interest repayments over the first 5 or 10 years. Interest payments may be fixed for a certain period and then converted into a variable rate. Some borrowers do not want to be exposed to variable interest rates and may negotiate what are termed 'drop lock loans', which allow the borrower to switch from a variable rate of interest once the rate reaches a certain level.

Corporate finance

As we have already discussed (see 'Property companies and the stock market'), there are various methods available of raising equity and debt finance from institutional investors via the stock exchange, which we will now examine briefly.

Equity finance

1. **New shares**   Companies may raise money by selling shares to investors in a floatation on the stock market or the unlisted securities market. The majority of new share issues are underwritten by financial institutions for a fee, and they will buy any shares not bought.

   Generally speaking there are two types of shares: ordinary and preference. An ordinary share is a share in the equity or, in other words, part-ownership of the company. Ordinary shareholders have voting rights and share in the risks and profits of the company. Profits after tax are distributed via dividends, usually half-yearly. Companies may also issue convertible preference shares at a fixed dividend, which, within a specific period, may be converted into ordinary shares. Preference shareholders rank above ordinary shareholders in entitlement to dividend payments. However, preference shareholders do not participate in any growth in the company profits and normally have no voting rights.

2. **Rights issues**   A company can raise additional capital by offering existing shareholders the right to purchase a number of additional shares in proportion to their existing shareholding, at a lower price. As with new issues a rights issue is normally underwritten. The net asset value (NAV) per share will be diluted. The ability of a company to raise capital via a rights issue will depend on stock market conditions, the state of the property market as measured via property share performance and the NAV per share. In the past some companies have been able to successfully raise capital on the stock market via rights issues as property share prices performed well, while only marginally diluting the NAV of their shares.

3. **Retained earnings**   One source of finance is the company's own resources generated by profits.

   However, some profit will need to be distributed to shareholders as dividends. How much of the profits is paid out as dividends or retained is up to the company to decide, but they must be aware of the interest of their shareholders in maintaining a reasonable dividend.

Debt finance

Debt finance instruments may be secured on specific property assets or the property assets of the company as a whole. Alternatively, they may be unsecured where investors have to rely on the financial strength and track record of the company.

1. **Bonds**   A bond is considered a relatively low-risk investment and often the return is also relatively low in comparison to other investment options. Effectively it is an 'I owe you' note, secured on a specific investment property or a completed and let development owned by the company. Investors in a bond receive interest on a regular basis (e.g. each year) and their initial investment is repaid at a specific date in the future. Bonds are securities that can be traded on the stock market. The interest payments (known as the coupon) can be structured to avoid the usual problem of rental income shortfall. With 'stepped interest' bonds the investor receives a low interest rate initially, which rises at each rent review. An alternative is a 'zero coupon' bond where no annual interest is paid but the investors are repaid on the redemption date at a premium. However, both of these types of bonds rely on rising property values, which does not always occur due to the cyclical nature of the real estate market.

2. **Debentures**   These are securities that can be traded on the stock market. Debentures are issued by companies to institutional investors whereby the institution effectively lends money at a rate of interest below market levels in return for a share in the company's potential growth. The money is typically lent long term, usually up to 30 years, at a fixed rate of interest and is secured upon the company's property assets. Normally, the security is specifically related to named properties, but sometimes provision is made to allow the company to substitute one property for another subject to agreement on valuation.

3. **Unsecured loan stock**   Property companies may issue unsecured loan stock (not secured on the assets of the company) to institutions at fixed rates of interest that, within a specific period, can be converted at the option of the institution into the ordinary shares of the company. However, to reflect the higher level of risk attached to the absence of a high level of security, the interest rate is also higher. The higher interest rate also covers instances when the lender is unable to recover all or part of the loan, mainly due to the lack of security.

Unitisation and securitisation

The property markets and financial markets have recently developed equity financing techniques to reduce the problem of the illiquidity of property investment. They are attempting to make property investment more comparable to other investments and overcome some of the inherent obstacles including lack of a central marketplace, indivisibility (i.e. either purchase the entire property or not at all), transparency (i.e. lack of information about the product) and illiquidity (i.e. to access the money tied up in the property would normally take months to release, and include advertising, negotiation and the contract stage). Another obstacle is that to finance large, single developments worth over £150 million with funding institutions is difficult for developers unless two or more funds become involved. At times this may mean that larger single properties are valued at a discount compared with smaller investments due to the limited number of potential purchasers.

Many global markets have accepted that securitisation and unitisation of property investment is a means of broadening the demand for property beyond the existing financial institutions who are large enough to participate – for example, as a REIT listed on the stock market that smaller investors can buy part thereof. Simply explained, unitisation means the splitting up of ownership of a property or a portfolio of properties amongst several investors. Securitisation is a general term used to describe the creation of securities that can be traded on the stock market, e.g. shares, bonds, debentures and unit trusts. Therefore, the creation of securities is one way of achieving unitisation. Importantly, with each method the investor receives a return in exact proportion to their original investment.

It is worthwhile to discuss the background to various securitisation and unitisation techniques that have been introduced so far.

There have been various attempts at 'unitisation' of large properties, i.e. splitting the ownership of the property into small manageable chunks, allowing several investors to invest in the property. However, due to legal complications it has proved very difficult in practice to actually divide ownership of an individual property. Earlier attempts at the unitisation of individual properties have included SPOTS (Single Property Ownership Trusts) and PINCS (Property Income Certificates). SPOTS involve a trust owning a property and spreading the ownership among investors in the form of units similar to unit trusts; however, this method faltered due to tax problems. Nonetheless, the Inland Revenue could not be persuaded to allow the income from the property to pass to the unit holders without prior deduction of tax. PINCS involved a complex structure of companies and leases – a PINC is a security consisting of an income certificate, a contract to receive a share of income of the property after management costs and tax, and an ordinary share in the management company that manages the property. As a security, it was capable of being traded on the stock exchange, based on the concept that investors receive the benefits of ownership, in the form of a share in the income and capital growth, without owning the property direct. A PINC was not subject to income tax or capital gains tax. Although PINCS were ready for launch in 1986, they were abandoned in 1989 due to stock market conditions.

An earlier attempt at unitisation was undertaken via a company structure, such as where a company invests in a single property. In this case, Billingsgate City Securities invested in a single property, named Midland Montagu House, an office building in the City of London. It involved three different classes of security: deep discounted bonds, preference shares and ordinary shares. The preference shares were available to the public but did not prove popular. The issue was not successful and, as a result, the owner had to take over the majority of the issue. Single asset property companies (SAPCO) have an ordinary corporate structure and, therefore, have tax problems as the company attracts income and capital gains tax.

Previously there have been other ways of unitising a portfolio of properties via unit trusts aimed at small pension funds and private investors. Predominantly there are two main types of unit trust: authorized property unit trusts (PUTs) and unauthorised unit trusts. PUTs invest directly in property on behalf of their investors, which may include private investors. They are strictly regulated to ensure that they invest in a diversified portfolio of low-risk, prime income-producing property as part of an overall balanced portfolio including property securities. A PUT is treated as a company for the purposes of corporation and income tax, but is exempt from capital gains tax. On the other hand, unauthorised unit trusts are unregulated trusts investing in a mixed or specialised portfolio of properties. They are attractive to tax exempt financial institutions such as small pension funds and charities, which lack the funds to invest in property directly. Where all the investors in the trust are tax exempt then the trust is exempt from capital gains tax. Unit trusts are managed by a committee of trustees elected by the unit-holders (investors) under the terms of the trust deed.

4.4 The future

The property development industry is part of the larger real estate market, which is subject to changing supply and demand levels resulting in often clearly defined property cycles. History has shown that poor timing by a property developer can result in completion at the bottom of the cycle where rents are low and demand is scarce. Since the last property bust in the mid-1990s the lenders, primarily in the form of banks, have taken a vested interest in the projects they are lending 'their money' on. Accordingly, most lenders now take an active role in understanding the dynamics of the property market and the likelihood that the development will reach its full potential. Thus a borrower must provide detailed market evidence and projections in order to convince the lender that the profit levels will actually be achieved.

It is critical for a borrower to undertake the role of a lender in order to borrow money on suitable terms. Simply explained the lender is seeking at all times to decrease their exposure to risk, primarily in the form of either property-specific (unsystematic) risk or market (systematic) risk. In return for accepting a perceived higher exposure to risk, the lender will charge a higher interest rate commensurate to the level of risk. Whilst many forms of risk are unavoidable, such as risk due to the time needed for development, often a proportion of the risk can be reduced by the borrower – for example, pre-letting or pre-selling a development will remove risk associated with both the final rent/sale price as well as the likelihood that the property will remain void after completion.

At any given time there will be myriad financially strong companies; new players embarking upon development schemes on the basis of pre-lets and, in some cases, speculative schemes. There are a large number of established and emerging lending products that are ideally suited for each project and the lenders are actively seeking to lend money. At the same time there are property developments that may not succeed and many lenders will avoid these projects, forcing the property developers to rethink their proposal or even to proceed with the project in the current property market climate. Many vacant sites are testament to property developers waiting for the optimal time to initiate a proposal or reapproach a lender when the market is on the rise.

Property developers have a wide variety of lenders and associated products available to them, although the market is constantly changing and adjusting to its own supply and demand forces. In the future the property developer must be constantly seeking to keep abreast of changes in the lending market, especially when considering changes in taxation and legislation. Only then will the property developer be able to develop a competitive product to realistically compete with other property developments, especially considering that the interest and borrowing costs are such a large proportion of the overall property development. Overall, it can be argued that banks will remain as short to medium-term debt financiers, although securitisation and unitisation are becoming more readily accepted in both the property and equity markets. The rapidly growing acceptance of REITs throughout the world should continue, especially since this investment medium has the ability to overcome many of the negative benefits associated with direct property investment and is ideal for larger pension funds.

Reflective summary

There are a variety of sources and methods of financing property development both in the short and the long term. The choice and availability of funding will depend on the nature of the scheme, how much risk the developer wishes to share, and the confidence of both the financial institutions and the banks in relation to the underlying economic conditions at any particular time. The most secure route from the developer's point of view, provided the development is prime, is forward-funding with a financial institution, a method that combines both short-term and long-term funding. However, if the developer wishes to retain flexibility, either wanting to retain the investment or sell it when market conditions are favourable, then debt finance is more appropriate in the short term. The terms and method of debt financing will depend on the financial strength of the developer and the value of the security being offered. A pre-let scheme being carried out by a financially strong developer represents the best proposition. The greater the risk the less likely the developer will be able to obtain debt finance on favourable terms, unless the developer either contributes its own capital or shares the eventual profits. If debt finance is used then both the developer and financier must have regard to the availability of long-term finance and the requirements of property investors. Property has to compete with other forms of investment that offer more liquidity to the investor, so funding and valuation techniques have to be developed to improve the attractiveness of property as an investment.

CHAPTER

Planning

5.1 Introduction

Since the last edition of this book, there have been changes to the organisation and system of planning in the UK. The system remains one of the most sophisticated in the world; however it is now more influenced by the European Union (EU). The main area of EU influence is in environmental control, which is dealt with in Chapter 9. It is necessary for developers to obtain a specific planning consent for virtually every project. The only exceptions are those minor projects covered by General Development Orders, though there a large number of these. It is beyond the scope of this book to cover in detail all aspects of the planning system and readers are referred to other texts for further reading. The aim of this chapter is to provide an appreciation of the principles involved in the planning process since it is a vital part of development practice. The planning process will determine the type of development allowed on any site and thus influence the value of the development. Criticisms are that the planning system and process adds uncertainty to development in the form of delays, potential cost increases and, at times, unpredictable decisions. These are issues that concern developers; however, the past decade has seen an ongoing series of changes and review culminating in the UK government's Barker Review of Land Use Planning, published in December 2006.

5.2 Planning and the environment

The Barker Review of Land Use Planning stated that the aims of planning policy and procedures are to deliver economic growth and prosperity alongside sustainable development goals and this reflects a changing emphasis that commenced in the mid-1980s. (See Chapter 9 for further details on the environmental aspects of the current UK planning system.) Broadly speaking, it is argued that the consideration of sustainable development goals has lead to a more rigorous study of development proposals and wider involvement of third-party consultees but also frequent delays in the consideration of planning applications.

As a consequence of the debates held at the 1992 Earth Summit in Rio de Janeiro, and the United Nations Bruntland Commission, the UK government is committed to the concept of 'sustainable development', i.e. 'meeting the needs of the present without compromising the ability of future generations to meet their own needs' (United Nations, 1987). In 1988 the EU parliament issued a directive requiring the environmental impact of a large number of developments to be formally assessed, and the subsequent publication by the government of the Town and Country Planning (Assessment of Environmental Effects) Regulations set out specific requirements in this regard (see 'Environmental impact assessment'). Since the late 1980s there has been increased linkage between town planning and environmental law and regulations.

The key town planning legislation in England and Wales is the Town and Country Planning Act 1990 as amended by the 1991 Planning and Compensation Act, and the Planning (Listed Buildings and Conservation Areas) Act 1991. The most recent reforms to the planning system is the Planning and Compulsory Purchase Act 2004, which is intended to improve and speed up the planning system, with the goal of reforming the planning system. The key legislation in Scotland is the Town and Country Planning Act (Scotland) 1997. The details of the legislation and its interpretation are covered in more specialist books and the websites that are listed in the references at the end of the book comprise a broad summary of how the planning system operates and its implications for those involved in the property development process.

5.3 The planning system

A large number of government agencies and departments are involved in town and country planning especially following devolved government, introduced during the 1990s to Wales, Northern Ireland and Scotland. These bodies are: for Wales, the Transport Planning and Environment Group of the Welsh Assembly; for Scotland, the Scottish Executive Development Department (SEDD); for Northern Ireland, the Planning Service Executive Agency of the Department of the Environment for Northern Ireland (DoENI); and for England, until 2006, the Office for Deputy Prime Minister (ODPM). Communities and Local Government is now the government department responsible for building regulations and planning in England and was created in May 2006 by the Secretary of State (see http://www.communities.gov.uk for further details), replacing the ODPM. It has a key role in meeting the UK government's sustainability targets.

The planning system is overseen in the government by the Secretary of State for the Environment (and in Wales by the Secretary of State for Wales) who, in the context of the legislative framework provided by the Planning Acts, issues statements of policy and guidance (through ministerial Circulars, Planning Policy Guidance Notes and Planning Policy Statements) to local planning authorities on relevant planning matters, including their approach to applications of varying types. The Secretary of State also has the power to consider matters of strategic policy in county structure plans and Unitary Development Plans (UDPs), although this power is rarely used, and through an appointed inspectorate considers all planning appeals against the decisions of planning authorities. The Secretary of State also has the power to determine any application or planning appeal if it is of national or regional significance, via what is known as the 'call in' procedure.

The list of Planning Policy Guidance Notes (PPGs) and Planning Policy Statements (PPSs) issued by the Department of the Environment is available on the Communities and Local Government website. Currently PPGs are being replaced by more succinct PPSs. These statements of policy are rigorously scrutinised by those involved in the property development and planning process as they provide general guidance to all concerned (including local authorities) on how to approach planning applications for various forms of development ranging from housing, shopping and business parks to sports facilities and power stations.

In addition to the formal policy statements, the content of speeches and statements from relevant ministers, and particularly the Secretary of State for the Environment, are important considerations. An example of the importance of these statements was the 1994 controversy prompted by a speech by the Secretary of State relating to out-of-town retailing. As the issues that arise in the field of town planning (often brought to a head by applications for development) have become more politically sensitive, the role of politicians at all levels of government in the making of planning decisions cannot be overestimated.

Restructuring local government has become customary and the current system comprises a uniform two-tier system of local government established by legislation passed in 1963 for London, 1972 for Scotland, and in 1973 for the rest of England and Wales. General strategic policies are set out in the county structure plans by the county councils and the majority of the planning applications are determined by the district councils in the framework of structure plan policies, the policies of a local plan that the district councils may have prepared and the Secretary of State's general planning guidance referred to above. The upper tier (the Greater London Council and the metropolitan county councils of Greater Manchester, Merseyside, Tyne and Wear, West Midlands and West Yorkshire) were abolished leaving a unitary system of local government in these areas. Further restructuring took place in 1996 and a number of unitary authorities were established in parts of non-metropolitan England between 1995 and 1997. England has a varied structure of local government in which 115 areas have a unitary system and the remainder comprises two tiers, forming 34 counties and 238 districts (Cullingworth and Nadin, 2006).

In England, county councils are responsible for county matters: minerals, waste and assistance to the regional planning body in preparation of the regional spatial strategy, e.g. Essex County Council. The district councils are responsible for most local government planning functions except where there is a national park authority. Many provincial cities and a few rural areas have a single-tier structure with unitary councils, of which there are forty-six, responsible for all government planning functions. In the six metropolitan areas of Greater Manchester, Merseyside, Tyne and Wear, West Midlands, South Yorkshire and West Yorkshire there is a unitary structure with thirty-six metropolitan district councils responsible for all local government planning functions, e.g. Sheffield City Council. Note that the metropolitan county councils were abolished but the metropolitan counties still exist. In London there is a single-tier structure with thirty-two London boroughs, e.g. the London Borough of Tower Hamlets, including the Corporation of the City of London, responsible for all local government planning functions but working with the strategic policy of the Greater London Authority. In Wales there is a single tier of local government with twenty-two unitary councils responsible for all local government planning functions except in the areas of national parks. The town planning role of these authorities includes the preparation of a UDP that combines the strategic nature of a structure plan and the more site-specific nature of a local plan. The plans are prepared against the background of regional planning guidance from the Secretary of State for the Environment. The unitary authorities themselves determine all planning applications in the context of the UDP and the Secretary of State's general planning guidance.

Whilst this chapter will concentrate on the system as it applies generally in England and Wales (see comments on Scotland below), there are certain areas that are exceptions.

Simplified Planning Zones

Simplified Planning Zones (SPZs) are particular zones within a planning authority's area where certain planning restrictions may be relaxed and are set out in the Planning and Compulsory Purchase Act 2004.

England and Wales have a plan-led system, with a framework for the consideration of development proposals. Well-produced UDPs should embody a vision for the future of an area with which local people can identify. Equally, UDPs should provide for greater certainty and consistency in decision taking on individual planning applications The combination of the structure plan and local plan outside London and the former metropolitan counties, and the UDP in the unitary areas, is summarised by the phrase 'the development plan'. This description may also extend to more specialist plans covering minerals and waste disposal. The importance of the development plan was elevated by the Town and Country Planning Act 1990 not only because authorities are required to produce such plans but because of the increased prominence given to the policies of fully approved or adopted plans in the determination of applications. The key section of the 1990 Act is Section 54A, which states:

> In making any determination (of a planning application) under the Planning Acts, regard is to be had to the development plan; the determination shall be made in accordance with the plan unless material considerations indicate otherwise.

In these terms, the role of the development plan is a key one in the planning process. So far as those involved in property development are concerned, much more attention is now placed on the plans prepared by the various authorities. This manifests itself not just in relation to studying the relevant policies within the plans but more and more in the involvement in the preparation of the plans themselves through the making of representations at the appropriate time in an effort to influence their content (see below).

In the development plan process, the county council's role is that of strategic policy maker in the structure plan, a document that sets out general policies relating to a county, normally over 10–15 years. This would include the number of dwellings to be constructed in each district in the county, the amount of land required for employment-related development, the amount of new retail floor space required, those areas that should be specified for environmental protection and what new roads are proposed. The county council's proposed policies within a structure plan are the subject of several stages of public consultation (see Figure 5.1) and a public inquiry (the

1.	Initial consideration of policy issues statistics (e.g. household formation) and research by county council.
2.	Publication of initial consultation draft plan.
3.	Period for public comment and representations on consultation draft plan.
4.	County council considerations of representations and amendments to consultation plan.
5.	Publication of submitted draft structure plan for future presentation to the Secretary of State for the Environment.
6.	Period for public consultations and representations.
7.	Selected parties invited to attend EIP.
8.	EIP held with panel appointed by SOS presiding: discussion of issues identified by the panel as being of key significance.
9.	Panel discussion at EIP of representations made at stage 6: production of report of recommendations to county council relating to changes in policy deemed appropriate.
10.	County council considers changes to submitted structure plan resulting from panels report and publishes proposed modifications to submitted plan.
11.	Further opportunity for representations to be made.
12.	County council considers representation and then approves the plan.

* Note the SOS may decide at certain stages to formally consider and approve the plan.

Figure 5.1   Structure plan approval process

Examination in Public (EIP)) is held where invited interested parties can make representations seeking to change the county council's intended strategy. The EIP is presided over by a 'panel' appointed by the Secretary of State who considers all the arguments involving the county council concerned, the district councils within the county, developers, environmental groups and other local organisations. The panel report with recommendations to the county council who decide whether to amend the plan. Whilst the Secretary of State retains powers to intervene, in practice these are exercised rarely. Developers are often critical of this process as they see the county council as being in a position to ignore any panel recommendations they do not like. Those who favour the system argue that it allows for important strategic decisions to be made at county rather than national level.

Counties and most non-metropolitan unitary districts are covered by structure plans, in which the county, national park, or unitary authority set out strategic policies as a framework for local planning, and local plans. District councils and national park authorities set out more detailed policies to guide development in their areas, including proposals for specific sites. Structure plans may be prepared on a joint basis between two or more authorities (e.g. a county and a unitary authority or a national park). County, national park and some unitary authorities also prepare minerals and waste local plans. In London and the metropolitan areas, and in a few non-metropolitan unitary areas, councils produce UDPs, which combine the functions of structure and local plans and include minerals and waste policies.

Structure plans broadly set out the general policies and proposals of strategic importance for the development and use of land in an area, taking account of national and regional planning policies. Structure plans do not contain detailed policies or site specific proposals to be used for development control, but focus on providing a strategic framework within which detailed policies can be framed in local plans. Structure plans may include policies on:

- housing, including figures for additional housing requirements in each district, and targets for development on previously developed sites;
- green belts;
- the conservation and improvement of the natural and built environment;
- the economy of area, including major industrial, business, retail and other employment-generating and wealth-creating development;
- a transport and land use strategy and the provision of strategic transport facilities including, highways, railways and other infrastructure requirements;
- mineral working (including disposal of mineral waste) and protection of mineral resources;
- waste treatment and disposal, land reclamation and reuse;
- tourism, leisure, sport and recreation;
- energy generation, including renewable energy.

Once the structure plan is approved, the county council's role is to ensure that the plan is regularly monitored and reviewed, where necessary. In terms of the submission of planning applications on a day-to-day basis, the county council mainly acts as a consultee to the district councils, with the exception of specific types of planning applications that it has the power to determine (e.g. those dealing with mineral extraction). However, it retains advisory powers relating to highway matters, which are important for many major schemes requiring planning permission. The district council's role is to consider and determine the vast majority of all planning applications submitted. They are also required to prepare, in the context of the county structure plan, local plans covering their district, where specific proposals are made and policies are set out for the control of development. An example of the relationship between a structure plan and a local plan is where a structure plan requires 5,000 dwellings to be constructed in a particular district over a 10-year period to meet strategic needs and the local plan allocates specific sites for development in the district to enable the requirement to be met.

The 1990 Act requires district councils and unitary authorities to prepare local plans or UDPs (see also http://www.communities.gov.uk). The preparation of a local plan follows a series of well-defined stages that allow for public consultation on emerging policies. These are set out in Figure 5.2. The development industry regularly involves itself in each stage and is usually represented when the objections to the plans made during their deposit stage (i.e. the deposit draft plan) are considered at a public inquiry.

1.	Initial consideration of issues by local planning authority including reference to the strategic requirements of structure plan (or regional guidance for unitary authorities).
2.	Publication of consultation draft local plan.
3.	Initial opportunity for representations to be made (including county council to agree plan conforms to structure plan).
4.	Authority considers changes to plan resulting from representations.
5.	Publication of deposit draft local plan for 6 week formal public consultation.
6.	Six week period for formal objections (note only such objection as those which can be considered in stage 8 below).
7.	Authority considers changes to the plan resulting from objections (possibly by negotiations with others).
8.	Local plan inquiry into objections held with inspector presiding.
9.	Inspector prepares a report of recommendations of change to deposit draft plan arising from consideration of evidence at the local plan inquiry.
10.	Authority considers changes to plan arising from inspector's recommendations and advertises modifications to plan they decide to make.
11.	Period for representations to modifications.
12.	Authority considers stage 11 and proceeds to approve the plan.

Figure 5.2   Local plan/unitary development plan adoption process

The public inquiry is presided over by an inspector who listens to evidence objecting to and arguments for maintaining the policy of the deposit version of the plan. The inspector does not have the power to impose their view on the planning authority and a report is compiled making recommendations to the authority on whether each objection should result in the deposit draft plan being changed. The planning authority then decides whether to act upon the recommendations or not. If they do not, they have to justify their decision but they are not obliged to follow the inspector's views. After making appropriate modifications to the plan, and a further consultation period, the local plan or UDP can be 'adopted' thereby giving it the prominence referred to in Section 54A of the 1990 Act.

The lack of obligation to follow inspectors' recommendations frustrates developers, particularly if they have been successful in persuading the inspector to recommend that changes are made to a deposit draft plan after hearing evidence at the public inquiry. The contrary view is that it is right for decisions of this nature to be made by local/unitary authorities on the basis that local democracy is seen to take precedence on planning matters.

As set out in Figures 5.1 and 5.2, whilst there are ample opportunities for involvement in the evolution of structure plans, local plans and UDPs, the time taken for the processes to run their course is considerable. It is not unusual for the entire structure plan process to take 2 years to complete from the initial consultation to the final approval, whilst a local plan or UDP can often take over 3 years and this is a key recommendation of the Barker Review to streamline the planning system and to make it more transparent. See Table 5.1 for an example UDP timetable.

Table 5.1 An example UDP timetable

Period of presentation	February 2001
Consideration of representations	April–October 2001
Publication of deposit draft plan	November 2001
Public inquiry into objections received	
To deposit plan	October 2002–January 2003
Receipt of inspector's report	February 2004
Period of modifications to the plan following inspector's report and additional consultation	July–September 2004
Final adoption of plan	December 2004
Total time elapsed	3.75 years

The delay, uncertainty and cost that is a result of the process is a problem to developers, particularly as it is commonplace for there to be slippage in the local authority's envisaged timetable. The Barker Review noted that 69 per cent of organisations were dissatisfied with progress made by local planning authorities to improve the system (Barker Review, 2006). The government has made it clear in its PPG that the determination of planning applications must not be delayed pending the approval or adoption of development plans and that only if a proposal is so significant that it may prejudice the decision-making process should it be regarded as 'premature' to the finalisation of the plan. Similarly, government policy allows for account to be taken of plans as they progress through the various stages with more weight being attached to their policies the further along the process they proceed.

However, particularly in the areas of the out-of-town retail, business parks and residential development where greenfield sites are proposed for release by developers, it is often the case that such proposals are having to be pursued through the development plan process rather than through immediate planning applications. To that extent, the timing of preparation of a development plan has become, since the 1990 Act, a considerable factor in the site acquisition process for developers. Following the introduction of the Planning and Compulsory Purchase Act 2004, old-style development plans are being replaced by a new system of Local Development Frameworks (LDFs). The background to the new approach is set out in Planning Policy Statement 12: Local Development Frameworks, produced by the ODPM. Part II of the Planning and Compulsory Purchase Act 2004 provides a new development plan system based on Regional Spatial Strategies (which replace Structure Plans) and Local Development Frameworks (which replace Local Plans and Unitary Development Plans) in England. Local planning authorities must prepare a Local Development Framework that will comprise a folder of Local Development Documents (LDDs) for delivering the spatial strategy for the area (as opposed to the old single plan covering the whole of the authority's area). LDDs will comprise Development Plan Documents and Supplementary Planning Documents, which expand polices set out in development plan documents or provide additional detail. The LDF will also include a Statement of Community Involvement, the Local Development Scheme (which sets out the programme for the production of LDDs) and the Annual Monitoring Report.

The Scottish planning systems operates in a similar way to England and Wales in relation to the submission of planning applications, the preparation of development plans, the powers of planning control and the general appeals process. In April 1996 the two-tier system was abolished and replaced with a single tier or unitary system. The unitary authorities are largely based on the old 'districts' with certain exceptions, for example the former Highland Region is a unitary authority. In terms of planning responsibilities, there is no obligation on the new authorities to prepare unitary plans; instead, the authorities were instructed to have comprehensive local plan coverage for their areas (in one or more plans per authority). In structure plan terms, groups of unitary authorities co-operate (via joint planning boards) in the preparation of a plan for their area, which need not follow the boundaries of the former regions. This led to some confusion, and concern that potential joint planning boards might not have the expertise to produce such plans. Whilst the day-to-day operation of the system in relation to development control matters remained largely unchanged, the development plan function and responsibilities were substantially altered by the new system.

Discussion points

* What are the prevailing influences on the planning system in England and Wales?
* What are the major frustrations to property developers with the existing system?

5.4 What is development?

The statutory definition of development as set out in Section 55 (i) of the Town and Country Planning Act 1990 is:

> the carrying out of building, engineering, mining or other operations in, on, over or under land, or
> the making of any material change of use in any buildings or other land.

Broadly speaking, development might be thought of in two categories: one being the physical operations such as building or engineering works and the second being the making of a 'material change of use'. A book about property development tends to focus on the physical activities, but the question of use is important. Indeed, it is possible to argue that planning control is basically one of land use, because once the use of land has been determined, the question of precisely what is built is a matter of detail, albeit most important.

For the purposes of the planning control over changes of use in land and buildings the key document is the Town and Country Planning (Use Classes) Order 1987. The Order divides various land uses into classes, each separate class consisting of a group of very similar uses. Normally a change from one use to another within the same use class will not constitute development, whereas a change from a use that falls within one class to a use in another will (with certain exceptions) constitute development for which planning consent will be necessary. In practice, buildings are often suitable for a variety of different uses and a change from one use to another might involve a change of use for which planning consent will be necessary. The value of such a building might be substantially affected by the possibility of being able to obtain the necessary planning consent to change from one use to another. For example, a warehouse (Class B8 – storage and distribution) on the edge of town may be suitable for retail uses like do-it-yourself (Class A1 – shops and retail outlets). Appendix A of the Order contains a summary of the use classes.

The complexity of the uses to which land and buildings can be put, linked to changing trends in the types of usage, means the effectiveness and relevance of the Use Classes Order is constantly under review. A significant change was to incorporate offices (other than those in Class A2 providing financial and professional services), research and development and light industrial uses within one class, B1 Business, provided that the use can be carried out in any residential area 'without detriment to the amenity of the area by reason of noise, vibration, smell, fumes, smoke, soot, ash, dust or grit'. This change was made in recognition of the needs of modern industry, particularly 'high-tech' industries that, to varying degrees, often carry out office, research and development and light industrial uses within one building. During the 1980s development boom it had a dramatic effect on industrial land values as landowners expected office values for industrial sites. Many local authorities resisted the change but were not supported by the inspectorate at planning appeals. The Use Classes Order is not comprehensive in covering all uses and there are commonly disputes between applicants and local authorities over what are known as *sui generis* uses, i.e. uses not within any defined class in the Use Classes Order. Currently motor vehicle showrooms are classified as *sui generis*, but with permitted development rights allowing a change of use to A1 to be removed. This restriction is intended to prevent large, and often out-of-town, showrooms becoming retail shopping outlets without planning permission. It has been stated, however, that these changes could affect the actual value of motor vehicle showroom premises. In many cases, this is a major asset of the business against which funding is often obtained.

While the description of the physical works that need planning consent appears straightforward, circumstances will arise in which there is room for argument over the interpretation of the statutory provisions and a considerable volume of case law has been established as to what constitutes building or engineering, mining or other operations for the purpose of planning control. Arguments are often concerned with the placing of mobile structures on land or the installation of plant and machinery in a building or physical works carried out inside a building (which in no way affect the external appearance), which do not need planning consent, assuming that in both cases no material change of use results or that the building is listed.

5.5 e-Planning

One of the many changes since the previous edition of this book has been the take-up of information technology in the workplace. The e-Planning Team was set up as part of Communities and Local Government to support the work of the e-Planning Programme Board and to advance an e-Planning vision, which is to provide a world class e-planning service. An online service for the public and professionals has been established, where explanatory guides, application forms and full online submission and tracking of applications is possible. See the planning portal at www.planningportal.gov.uk for further information and also the Planning Inspectorate website at www.planning-inspectorate.gov.uk for details about planning appeals and full decisions.

5.6 The planning application

Anyone carrying out a development for which planning consent is required must normally apply to the local planning authority (either the district council or unitary authority) for consent. It is prudent to check that planning consent is necessary and a development will not require planning permission if it falls within the provisions of the General Development Orders (GDOs). GDOs have been extended to remove a wide range of developments from planning control and the need for planning consent. Where it is not clear if planning consent is required, the planning authority can be approached for informal advice. It is possible to formally establish whether a proposed use of land or buildings or operation is lawful by applying to the local planning authority for a formal 'certificate of lawfulness', to determine whether the proposals would constitute development for which a planning consent is required under Section 192 of the Planning and Compensation Act 1991.

The local planning authority must give a formal decision within 8 weeks and there is a right of appeal against the decision.

Anyone may apply for planning consent in respect of any property. It is not necessary for the applicant to have any legal or financial interest. However, if the applicant is not the owner of the property, a notice must be served on the freeholder, on any lessee with an unexpired term of at least 7 years still to run and on any occupier of agricultural property, advising them of the application for planning consent. The planning authority has a set of forms upon which these notices must be served, and the applicant must advise the planning authority of the names and addresses of the people on whom they have been served. In certain instances of development, e.g. buildings for various types of public entertainment or buildings exceeding certain heights, these are publicised by the local planning authority putting a notice in the local press; the notice makes it clear where any member of the public might inspect the plans showing the proposed development. Where the development might have a significant effect on neighbouring property, the authority is obliged to draw the attention of neighbours to the application. Any owner or lessee or occupier of agricultural land upon whom notice is served, or any other member of the general public, has the right to make representations to the local authority. There are several types of application that can be submitted to a local planning authority dependent upon the type of development that is proposed and the location of the site concerned. These principal types of application are as follows:

1. **Outline planning application**   An outline application seeks to establish the principle of a particular form of development outside Conservation Areas, without the need to deal with the matters of the siting of the buildings, their design, external appearance, landscaping or the means of access into a site. These 'reserved matters' can either in whole or in part be left for the future submission and determination of the planning authority in the event of outline permission being granted. A typical example of such an application would be for the residential development of a greenfield site on the edge of a settlement.

2. **Full or detailed application**   A detailed application will seek not only to establish a land use principle but approval for all the reserved matters listed above. The application is comprehensive and in the case of a residential development wovuld include not only information about the location of the site but also the layout of houses and roads, the design of the dwellings themselves, the principal landscaping proposals and all the relevant technical information. As the design of new buildings

and other reserved matters are particularly important considerations in Conservation Areas (see below) it is mandatory to submit full applications in such areas.

3. **Changes of use**   Legally an application for the change of use of land or, more commonly, buildings is regarded as a full application rather than an outline. This requires an applicant to submit full details of their proposals to the planning authority. Changes of use applications are normally (though not always) related to the uses defined in the Use Classes Order where such permission is required to change from one category of use to another. A typical example would be the change of use from a Class A1 retail shop on a high street to a Class A2 professional/ financial use such as a building society.

4. **Applications for listed building consent or Conservation Area consent to demolish**   All proposals to demolish buildings in Conservation Areas require specific permission and thus proposals for development in a Conservation Area must also, if demolition is envisaged, be accompanied by a separate application for Conservation Area consent to demolish. Similarly, any changes proposed to listed buildings, i.e. those specifically identified by the planning authority and English Heritage as being worthy of protection, must be the subject of specific applications.

A developer wishing to carry out building or engineering works should consider applying for an 'outline' consent, which establishes the principle of development proposed. The decision whether to submit an outline or detailed application will depend upon:

- the location of the site
- the issues involved
- the nature of the developer's legal interest in the site

For example, the developer may only have an option to acquire the freehold of the site or a conditional contract that is triggered by the granting of a satisfactory planning consent. Preparation of design drawings for a large building project can be costly and the submission of an outline application may save a good deal of time and trouble. An outline application must give sufficient information to describe adequately the type, size and form of the proposed development. The local planning authority will reserve for subsequent approval the reserved matters and will attach conditions to an outline planning consent to cover such issues. A typical condition (which will be one of many in an outline permission) would be as follows:

Before any development is commenced, detailed plans, drawings and particulars of the layout, siting, design and external appearance of the proposed development and means of access thereto together with landscaping and screen walls and fences shall be submitted to and approved by the local planning authority and the development shall be carried out in accordance therewith.

Fees are payable to the local planning authority in respect of applications, and the appropriate fee calculated in accordance with scales prescribed under Section 303 of the Town and Country Planning Act 1990 (usually reviewed annually) must be paid at the time the application is submitted. The fees are charged on a sliding scale depending on the size and nature of the development. As noted, the planning authority is required to determine each planning application within an 8-week statutory period, which runs from the date the application is registered by the determining authority together with the correct fee. If no fee or an incorrect fee is deposited with the planning application, the 8-week period will not start to run until the correct fee is paid. However, as of June 2004 only around half of the local authority's were processing 65 per cent of applications for minor projects within the 8-week time frame (Communities and Local Government, 2007).

Planning applications must be made on forms provided by the local planning authority and are available online. The forms are self-explanatory but many applications are delayed because applicants either do not complete the forms accurately or fail to provide all the information required. When the application form is submitted, the planning officer (a qualified person employed by the local planning authority who will handle all matters arising from the application) should be asked to confirm that the form gives all the information required and that no additional information is needed from the applicant and also that no supporting documents are necessary. The application will take its turn in a queue to be processed and checked; the site will be inspected by planning authority staff and there is commonly consultation with other organisations, such as the highway authority or the water authority on site restrictions, or with amenity societies in areas of special environmental interest. The planning authority usually consults other appropriate authorities before giving planning permission. For example, the Highway Authority will be consulted on the design of any new roads that are ultimately intended for adoption as public highways, or the local Environmental Health Officer might be consulted on potential noise or pollution problems, or the Health and Safety Executive might be consulted on 'hazardous' or potentially explosive processes that may be proposed in the development.

The process of making a planning application is summarised in Figure 5.3.

There are various matters upon which it is important that developers should satisfy themselves by means of direct discussions with the appropriate authorities. The more important of these are the adoption of roads and footpaths by the Highway Authority where it is appropriate for them to be adopted; the acceptance by the local authority of public open space provided within any development; and the necessary approvals under the building regulations, fire regulations or any other statutory provisions that relate to the development in question, e.g. the Factories Acts, the Shop Acts and Public Health Acts. Developers will agree direct with the Highway Authority the design and specification for all new roads and footpaths and enter into the necessary adoption agreement. They will reach a similar agreement with

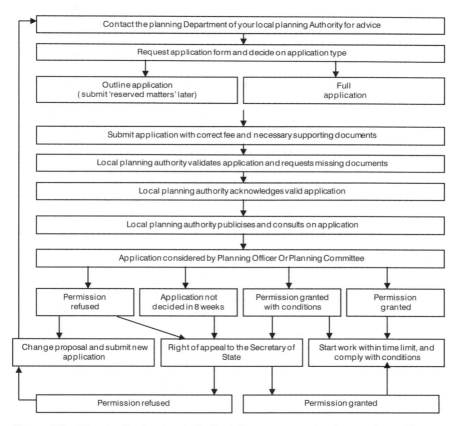

Figure 5.3   The planning process in England (Source: www.planningportal.gov.uk)

the appropriate water company in respect of sewers. A planning permission in no way constitutes an agreement for adoption.

During this process the staff of the planning authority may discuss with the applicant points that need clarification or problems that are revealed. At this consultation stage, it is often possible to make modest corrections or alterations in the application to avoid such problems, but more radical alterations might require another round of consultations, or a new application. At the time of submitting the application, it is useful to ascertain the date of the council's planning committee at which the matter will be considered and to check at the appropriate time that it has been placed on the committee agenda.

As discussed above the local planning authority pays particular regard to the provisions of any approved development plan as set out in Section 54A of the 1990 Act and any other material considerations in reaching its decision. However, if it is proposed to grant a planning permission that will constitute a substantial departure from the development plan, the local planning authority must advertise the planning application and give opportunity for objections to be made. The Secretary of State must be advised of the proposals and may decide to determine the application, i.e. to arrange a public inquiry to hear the arguments for and against a particular proposal. In the absence of any intervention by the Secretary of State in such cases, the local planning authority may grant permission. The Secretary of State has the discretion to 'call in' an application, so that it can be considered by Communities and Local Government. In practice, the Secretary of State intervenes when it is considered that the application is of national or regional significance, for example, shopping developments over 100,000 square feet in size or proposed new settlements.

Discussion points

- Under what circumstances might it be prudent for a developer to make an application for outline planning permission?
- What other local authority departments consider the impact of the planning application, and why is this?

5.7 Environmental impact assessment

Since the European Community Directive 85/337 and the Town and Country Planning (Assessment of Environmental Effects) Regulations 1988 introduced the need for an environmental impact assessment (EIA) in certain instances, a developer now may be required to produce an environmental statement (ES) as part of a planning application.

This statement should examine the impact of a development proposal on the environment in its widest sense. There is a comprehensive list of factors that can be assessed dependent upon the type of development proposed. Typically this can include analysis of the following aspects:

1. impact on the landscape
2. visual impact
3. ecology
4. geology
5. archaeology
6. air and water pollution
7. contamination of land
8. noise pollution
9. wildlife conservation
10. agriculture

Some projects, such as oil refineries, motorways and major power stations require assessment in every case. There are about eighty types of developments listed in the regulations that are subject to assessment where they are judged likely to give rise to significant environmental effects, for example, mineral extraction and major infrastructure projects such as roads and harbours. A developer may apply to the local planning authority for an opinion on the need for an environmental impact assessment prior to a formal planning application. A developer has a right of appeal if they disagree with a local authority's request for an environmental statement. The Secretary of State may make a direction that an environmental impact assessment is required or in certain cases the developer may voluntarily prepare an assessment to allay any anticipated fears on environmental matters. The main difference to the normal planning process where an environmental impact assessment is required is that there are additional publicity requirements and an extended period of 16 weeks for the planning authority to determine the application.

The role of environmental considerations in the planning process has increased greatly since 1988. This has taken the form of PPS1 published February 2005, which sets out the government's overarching planning policies on the delivery of sustainable development through the planning system and PPG 13, which advises that the location of new development should take account of the need to reduce pollution generally and carbon dioxide emissions in particular.

It is important to consult the relevant planning officer of the district council or appropriate authority at the outset, before any formal planning application for a reasonably sized development is submitted – often a great deal of wasted time and effort can be saved in this way. Through early consultations it might be possible to agree the development proposals in principle with the planning officer. They do not always do so, but planning authorities have the power to delegate various powers of decision to their planning officer, thus applicants should always ascertain whether the planning officer is authorised to issue the necessary planning consent or whether a recommendation will be made to the planning committee. The planning authority aim to give a formal decision within 8 weeks of the receipt of the application or, such longer period as the applicant may agree. In practice any substantial application takes longer than the statutory period to reach a decision. If consent is then refused or granted subject to conditions to which the applicant has objection, an appeal may be made to the Secretary of State for the Environment. If no decision is given within the prescribed 8-week time limit, the applicant can assume that planning consent has been refused and an appeal made to the Secretary of State accordingly. This right of appeal is not universally applied by developers if they consider further negotiations beyond the 8-week period will result in planning permission being obtained.

Even though consultation with the planning officer does not guarantee agreement that a proposal is acceptable, applicants should at least thoroughly understand the views of the planning authority, and their attention will normally be drawn to all the relevant development plans and other policies, so that they are in a better position to consider how their application may be approached.

Some planning authorities appoint advisory committees to comment on certain types of development. Most commonly, architectural advisory panels are set up to give advice on the architectural merits of proposed developments that are to take place in sensitive areas. Planning authorities must also consult local residents' organisations or amenity societies of various kinds. Amenity societies are normally consulted when development is proposed within Conservation Areas and the local planning authority has a duty to advertise the receipt of planning applications for developments in those areas. A local Conservation Area committee might exist and its comments will carry weight. The planning officer should be able to indicate whether it is possible or advisable to consult such committees or voluntary organisations at the informal pre-application stage.

With environmental issues now very high on the political agenda, it is becoming increasingly important for developers to consult with the relevant pressure groups at an early stage. Although planning authorities are advised by the government not to exercise detailed control over design (except in Conservation Areas), such issues are commonly disputed. This is because matters of design are subjective and capable of being interpreted in several differing ways.

The particular sensitivity that is associated with applications for development within defined Conservation Areas or affecting listed buildings arises because of the special protection afforded to such areas and buildings by the planning legislation. Planning authorities have powers to define Conservation Areas, i.e. areas of special architectural or historic interest that it is desirable to preserve or enhance, and this process can take place in conjunction with the preparation of a local plan or UDP or in its own right. Once defined, all planning applications for development within the area must be detailed and, as set out above, separate applications must be made for any demolition. Government policy in PPG15 is that all proposals for development within the Conservation Area must preserve or enhance its character and there is commonly much debate between planning authorities, conservationists and developers over the design merits of schemes in Conservation Areas in the context of government policy.

If the development proposals entail the demolition or alteration of the character of buildings that are on the list of Buildings of Special Architectural or Historic Interest prepared by central government, i.e. listed buildings, then special procedures are necessary. As described above, a specific consent for the demolition or alteration of the listed building must be obtained before any work can be carried out. Applications for the necessary consent must be made to the local planning authority and advertised, so that any member of the public may make representations. There are a number of reputable groups with an interest in the protection of listed buildings (who are consulted by planning authorities, e.g. English Heritage) and proposals for alterations or demolitions of listed buildings. Developers may have to spend some considerable time discussing their proposals with interest groups, in an attempt to reconcile their commercial requirements and the sensitive refurbishment of a listed building.

If buildings appear under threat from development proposals, a local authority can issue a temporary listing known as a 'building preservation notice', which has immediate effect and lasts for a period of 6 months, to enable the Secretary of State to decide whether to formally list the building concerned. If the building is not subsequently listed there may be compensation payable to the developer for consequential loss.

Local planning authorities also consult archaeology interests, e.g. the Museum of London. In the late 1980s development boom significant archaeological finds were made (e.g. the Rose Theatre in Southwark, London, in 1989). If the site is of archaeological interest, then the developer must allow an archaeological dig or a watching brief by archaeologists to take place before development commences. The period of the dig and any compensation or contribution from the developer is often a matter of negotiation between the developer and the interest group. Under the Ancient Monuments and Archaeological Areas Act 1979, specific areas of archaeological importance can be entered and excavated for up to 6 months but often developers offer contributions to speed up this process. Here again, the development process may be delayed and, therefore, it is prudent for a developer to consult with archaeological groups. The general policy guidelines on archaeological matters are set out in PPG16 and should be considered prior to purchasing a site. Many archaeologists now act as private consultants to developers.

If any trees on site are protected by a tree preservation order, it is necessary to obtain consent before felling or lopping them. Normally the necessary application is made to the local planning authority but in certain cases the consent of the Forestry Commission is necessary.

The Advertisement Control Regulations enable planning authorities to exercise a very tight control over all external advertising (including site boards promoting any development; see Chapter 8). This is frequently a matter in respect of which the planning authority will have a policy document. Normally, the greatest weight is given to considerations of visual amenity and public safety in those cases where advertisements might possibly distract the attention of road users.

5.8 The democratic aspect

Some types of application (normally minor in nature) might be decided by the planning officers of a planning authority using powers delegated to them by the planning committee. Lesser items, such as house extensions or advertisements, might be dealt with quickly in this way. More important or substantial applications will be decided by the Planning Committee of elected representatives of the authority, who will meet regularly and normally have full powers to make a decision on the application on behalf of the council. The committee will be advised by planning officers whose advice is based upon their knowledge of the area and its problems and policies, and upon consultations they have had with such bodies as the highway and water authorities. Their professional advice (which takes the form of a report) will also be influenced by their knowledge of planning case law and how the council's policies operate.

Normally the planning committee will heed the advice of their professional officers and will also take note of any representations made by members of the public. However, it is important to note that the committee is not obliged to accept the officers' recommendation and they may make a decision against the recommendation. Development is often highly contentious and public comment might come to the committee from individuals or groups or in the form of petitions. The committee must consider the variety of advice and representation it receives and make a decision to approve or refuse the application or perhaps, in some cases, to refer it back to the applicant to seek a modification.

Public interest in planning is often strong when the existing status quo might be disturbed. The planning committee is a committee of elected councillors who, as part of the democratic political machinery, recognise the impact that the decisions they may make has on their political standing. It is important for developers to appreciate that, particularly where large, controversial schemes are being prepared, the political make-up of a specific planning authority, in practice, can be a major influence on the eventual decision made on the planning policy issues.

5.9 The planning consent

Planning permission may be qualified in various ways and we briefly consider here the more important of these. Permission may be granted subject to a variety of conditions, for a limited duration of time or for the personal benefit of certain people or organisations. With regard to time limits, there are two aspects to consider. Every detailed planning permission will lapse unless the development is commenced within 5 years from the date permission is granted or such other time as the planning authority stipulate. The Planning and Compensation Act includes provisions that change the duration of the planning permissions and consents. With outline permission, the necessary application for approval to the various reserved matters must be made within 3 years of the outline consent being granted. The development must commence within 5 years of the date of outline approval or within 2 years of the date of detailed planning approval, whichever is the later, subject to the imposition of any other specific time limit by the planning authority. Only a very limited amount of work on site (known as 'material operations') is necessary to prove that development has commenced to meet the time-limit requirements. The courts have decided that in order to show that a planning permission has been implemented; there must be a genuine intention to carry out the development concerned when a commencement is made on the site. This 'test' has caused developers problems, particularly if planning permissions are obtained at the height of a property boom and are not implemented and are then due to lapse during an economic slump. Developers must choose either to renew permissions (which require renegotiations with the planning authorities) or to let existing consents lapse. To deal with the evasion of the time-lapse provisions attached to planning permissions, local authorities have powers (Section 94 of the 1990 Act) to serve a completion notice, the effect of which is that planning permission will lapse unless the development is completed within a reasonable time. Completion notices must be approved by the Secretary of State and there are rights of objection for the people who are affected by them, which are considered by an inspector appointed by the Planning Inspectorate who is responsible to the Secretary of State for Communities and Local Government.

The second aspect of time limits is when the planning permission remains in force for a limited period and at the end of that period any buildings or works that have been erected must be removed and any use authorised by the permission must cease. At the end of the limited period of planning permission things will revert back to the state that existed before the permission was granted. On the expiry of limited period planning permission, it is open to the applicant to make a new application for the retention of the buildings or the continuance of the planning use. Planning permissions for limited periods are of limited value.

Conditions may be imposed limiting occupation to a particular type of occupier. These fall into two broad classes. There is the condition that limits occupation of a building to someone engaged in a particular trade or vocation – perhaps the best-known condition of this type is limiting the occupation of agricultural cottages to those engaged in agriculture. The other, and more restrictive type of condition, although rare, is one that limits occupation to a particular occupier personally. The ability to impose limitations on the type of occupiers sometimes enables a planning authority to grant a permission that otherwise it would not be prepared to consider.

The powers of planning authorities to impose conditions on a planning permission are set out by government advice in the Department of Communities and Local Government Circular 11/95. This Circular brings up to date Circular 01/85 on the use of planning conditions. It reflects guidance on the use of conditions in respect of transport, retail development, contaminated land, noise and affordable housing. Other matters covered include use of conditions for design and landscape, lorry routing, 'granny' annexes, staff accommodation, access for disabled people, holiday occupancy and nature conservation. The Circular also takes account of court decisions and includes an expanded appendix with model conditions. It records the policy that conditions should only be imposed where they are necessary, relevant to planning, relevant to the development to be permitted, enforceable, precise and reasonable in all other respects. A typical example might be a condition requiring landscaping proposed as part of a scheme to be carried out in the next planting season available and maintained thereafter to the satisfaction of the local planning authority. There is the same right of appeal against the imposition of a condition on a planning permission as there is against the refusal of planning consent.

Discussion point

What are the various qualifications that may be attached to planning permissions and how can these affect a developer's proposals?

5.10 Planning agreements or obligations

Whilst the planning authorities prepare the policies and plans to establish the planning framework for a particular area, this in itself does not bring about the implementation of the development plan. Implementation depends upon landowners and developers who initiate development proposals within the planning area. Having prepared the planning framework, the local planning authority normally waits for developers/landowners to appear and produce specific planning proposals and to make planning applications. In addition to the power to, where appropriate, attach conditions to the grant of permission it is possible for planning authorities to enter into legally binding agreements with developers that enable development proposals to come forward in circumstances where the authorities could not rely solely on their statutory powers of control, i.e. planning conditions. Planning agreements (referred to as planning obligations) between a developer and a planning authority may be made under the provisions of Section 12(1) of the 1991 Act, which substituted Section 106, 106A and 106b of the Town and Country Planning Act 1990. Section 106 introduced the concept of planning obligations that comprise both planning agreements and unilateral undertakings. Agreements might be made to phase the development of land to accord with the dates when various public services will become available or improved road access will be provided. Similarly, agreements might be made with regard to the provision of land within a comprehensive development area for public open space or amenity purposes. Where there is inadequate infrastructure, e.g. a lack of main sewers or adequate road access, it might be possible for a Section 106 agreement to be entered into requiring the developer to make financial contributions towards the cost of making available the infrastructure such that development may proceed. With regard to the provision of infrastructure, it should be remembered that agreements may be entered into with other authorities (relating to the adoption of roads and sewers etc.) as well as the local planning authority.

If developers feel that a local planning authority is attempting to exert undue pressure on them to enter into an agreement that will impose unduly onerous burdens, their remedy is to make a formal planning application and take the matter to appeal if planning consent is not granted. However, in many cases it will be to the advantage of developers to offer a planning obligation themselves if they consider that an appropriate contribution of a facility or infrastructure will enable the development to take place at a very much earlier date than would otherwise be the case. Since the 1991 Planning and Compensation Act, developers may offer a 'unilateral undertaking' to be bound by such an obligation. This can be considered as material by the planning authority in determining a planning application and at an appeal to the Secretary of State following a refusal of consent due to failure to agree terms on an agreement with the planning authority. Note that agreements bind landowners and successors in title and are registered on the legal title of the land affected.

There has been an increasing use of planning agreements, particularly involving local authorities with infrastructure burdens. Contributions towards infrastructure, whether they are given in cash or by way of sites for various public authority purposes, will obviously be reflected in the amount that a developer will be prepared to pay for the land for the development. However, often the developer has to assess the likelihood of a Section 106 agreement and its cost prior to site purchase, which adds uncertainty.

The extent to which a planning authority should, when considering a planning application, negotiate with a developer in order to obtain some material benefit for the community (referred to as 'planning obligation') has been a matter of controversy. Communities and Local Government Circular 05/05 provides guidance on the circumstances in which planning agreements are considered to be reasonable. The guidance confirms that planning agreements should only be sought where they are necessary to the granting of permission, relevant to planning and relevant to the development to be permitted. There are five tests of reasonableness:

- relevant to planning;
- necessary to make the proposed development acceptable in planning terms;
- directly related to the proposed development;

- fairly and reasonably related in scale and kind to the proposed development;
- reasonable in all other aspects.

Some feel that it is wrong for a planning authority to bargain to obtain a material benefit in return for a planning permission and that such a practice brings the planning system into disrepute. Many feel that developers should not be subjected to ad hoc demands from planning authorities, as they are seen as potentially dangerous precedents in the nature of local taxes on development. However, others argue that, in moderation and with common sense, planning gains can facilitate development in circumstances where the authority is not able to provide needed facilities and the planning gain makes a contribution to the welfare of the community in which the development takes place. Remember that the applicant has a right of appeal if planning consent is refused for any reason, and where a developer has submitted a unilateral obligation to enter into a Section 106 agreement on specific terms, the nature of the those terms and their appropriateness in the context of the Circular 16/91 tests is examined in detail.

5.11 Breaches of planning control

Local planning authorities have wide powers to ensure that no development that requires planning permission takes place, that no unauthorised uses are allowed to continue unless the planning position is regulated and that all development permitted is carried out in accordance with the conditions that the authority has imposed to the permission. These powers are included in the 1990 Act and summarised in their Policy Guidance Note.

Where development has allegedly occurred without permission the authorities are empowered to firstly require information concerning the development, the identity of the owner of the land involved and other pertinent matters to be divulged to them. They can then serve an enforcement notice on the landowner/developer that identifies the breach of planning control, the action required to remedy the breach and a time limit to carry out the necessary action. There is a right of appeal against the notice on one of seven grounds. The most commonly used ground is that planning permission should be granted for the development concerned and the lodging of an appeal on that ground is regarded by the Department of the Environment (who consider the appeal) as being deemed a planning application. Other grounds include stating that the alleged breach has not taken place and that the means required by the planning authority to rectify the situation are unreasonable.

Additional powers are available in the form of breach of condition notices, where authorities can require that conditions on a planning permission are complied with. There is no right of appeal against such a notice, as the notice is then pursued to the Magistrates' Court via a criminal prosecution, which can be defended. More immediate powers of action are contained in a stop notice, which literally requires a specific activity to cease, normally as a forerunner to enforcement action being taken. If an appeal is successful against enforcement action where a stop notice has been served, there is a compensation procedure.

5.12 Planning appeals

In the event of the local authority refusing planning permission for a proposed development, or in the event (subject to the discretion of the applicant) of the authority either taking longer than 8 weeks to determine the planning application or granting permission subject to a condition that aggrieves the applicant, there is a right of appeal to the Secretary of State for the Environment. To be valid, an appeal against a refusal of planning permission must be lodged within 6 months of the date of refusal. In deciding whether to lodge an appeal against a refusal of permission, an appellant must be satisfied that there is sufficient evidence to suggest that the local planning authority's reasons for refusal is inappropriate. Planning authorities are required by central government to ensure that reasons for refusal are sound and clear-cut, and the onus at the appeal is placed on local authorities to demonstrate that the refused proposals would cause demonstrable harm to interests of acknowledged importance. In assessing whether reasons for refusal are capable of being set aside on appeal, the appellant (and their advisers) would examine the relevant planning policies (particularly in the context of Section 54A of the 1990 Act) and any other material considerations.

Once a planning appeal has been lodged, jurisdiction of the case passes to the Department of the Environment's Inspectorate. This is a body of qualified inspectors appointed by the Secretary of State to determine planning appeals on his behalf or, in certain cases, to make recommendations to the Secretary of State to enable determination of the appeal himself. There are three methods of appeal that can be pursued by appellants.

1. **Written representations**   This is an exchange of written statements with the appellant's statements putting forward the case in support of the appeal being allowed and the local authority's statement in reply seeking to justify the reasons for refusal. Both statements are considered by the inspector appointed by the Planning Inspectorate, the government agency

responsible for determining appeals in England and Wales, who visits the site and determines the appeal on the basis of the written evidence and the site visit.

2. **Informal inquiry**   This takes the form of a meeting between the appellants and the local planning authority with the inspector acting as chair. Written statements of case must be submitted to the Department of Communities and Local Government in advance of the hearing. Having read the statements, the inspector usually identifies the issues that require discussion at the start of the hearing and these issues are aired between the parties. Following the hearing, a site visit is held and the inspector determines the appeal on the basis of the written statements, the hearing discussion and the site visit.

3. **Formal public inquiry**   This is a quasi-judicial inquiry where the appellant is represented by a barrister, solicitor or qualified person who presents the case for the appellant, calling expert witnesses to give evidence in support of the appeal. The planning authority is similarly represented, usually by the council's solicitor, who calls expert evidence supporting the refusal often from the council's planning officer. The witnesses are cross-examined by the opposing party, and the inspector, having listened to the evidence presented at the inquiry and following the site visit, then makes his decision or recommendations to the Secretary of State.

The three methods of appeal are the subject of detailed guidance on procedure through Department of the Environment Circular 05/00 Appeals. Department of the Environment Circular 8/93 provides for costs to be awarded against either party at an inquiry if there has been unreasonable behaviour. This definition normally applies to local planning authorities who have unreasonably withheld planning permission for proposals that were clearly acceptable or where grounds for refusal have been withdrawn at a late stage. Until relatively recently costs were infrequently awarded against appellants. However, the government has made it clear that, given its commitments to green belt policy and the development plan, in certain cases, therefore, where such proposals have been brought forward by developers to appeal (which are clearly contrary to an adopted development plan and where there are no material considerations suggesting the plan should not be followed), costs are to be awarded against the appellants.

With an increase in economic activity in the property boom of the late 1980s and again since the late 1990s, the number of planning applications that are refused, and the number of appeals, rises. Conversely, with a downturn or recession the level of activity declines. In addition, since the 1990 Act was passed and the emphasis was placed more on the development plan, developers have been advised to pursue schemes in the context of their involvement in the development plan process outlined earlier as a means of facilitating a favourable policy framework in which to submit an application. The number of planning appeals has declined in England and Wales to 22,689 in 2006–7 from 23,160 in 2004–5 (The Planning Inspectorate, 2007) but the success rate has remained the same at 34 per cent. This is a reflection of the government's intention for the planning system to be 'plan led'. The time taken for the process to run its course, given the need for evidence to be presented and considered by the inspectorate, is lengthy. It is normal for the public inquiry appeal route to take between 6 and 9 months to be completed from the date of lodging an appeal in receipt of a decision. In particularly busy local authorities where a backlog of appeals by public inquiry may have built up, the process is known to take much longer. The time taken for appeals by written representations and informal inquiries are somewhat shorter, but the delays that occur in the decision-making process are clearly a factor to consider when making a planning application that is likely to go to appeal. This has been a perennial problem with the planning system and one of the issues identified in the Barker Review (2006) that needs to be addressed. If the planning system is unable to respond quickly enough some development opportunities will be lost, as well as any associated economic growth potential.

5.13 The future

In that the planning system is slave to the political and legislative system, the future is difficult to predict. The development industry is generally dissatisfied with the delays and uncertainty that it experiences and sometimes the ad hoc nature of the decisions made, particularly where local politics holds sway over professional logic. Various proposals have been put forward to improve the system, the latest of which is the Barker Review of Land Use Planning in December 2006 (see Figure 5.4).

With the Labour Government's move to the centre of politics over the last decade, there has been an acceptance that planning plays a key role in delivering sustainability and economic growth for England and Wales. The current trend is for increased population growth in the England, increased globalisation, economic migration of peoples and increased need to adapt to and to mitigate the impacts of climate change. The key findings of Barker's review are for increased flexibility and responsiveness in the system, increased efficiency in the planning process and a more efficient use of land. As far as developers are concerned it appears that some of the historic problems remain in the system, those of delay for example, whilst newer and complex issues, like sustainability, have arisen and need to be addressed in property development.

Key Recommendations

- Streamlining policy and processes through reducing policy guidance, unifying consent regimes and reforming plan-making at the local level so that future development plan documents can be delivered in 18-24 months rather than three or more years;
- Updating national policy on planning for economic development (PPS4), to ensure that the benefits of development are fully taken into account in plan-making and decision-taking, with a more explicit role for market and price signals;
- Introducing a new system for dealing with major infrastructure projects, based around national Statements of Strategic Objectives and an independent Planning Commission to determine applications;
- Promoting a positive planning culture within the plan-led system so that when the plan is indeterminate, applications should be approved unless there is good reason to believe that the environmental, social and economic costs will exceed the respective benefits;
- In the context of the Lyons Inquiry into Local Government to consider enhancing fiscal incentives to ensure an efficient use of urban land, in particular reforming business rate relief for empty property, exploring the options for a charge on vacant and derelict previously developed land, and, separately consulting on reforms to Land Remediation Relief;
- Ensuring that new development beyond towns and cities occurs in the most sustainable way, by encouraging planning bodies to review their green belt boundaries and take a more positive approach to applications that will enhance the quality of their green belts;
- A more risk-based and proportionate approach to regulation, with a reduction in form-filling, including the introduction of new proportionality thresholds, to reduce the transaction costs for business and to increase the speed of decision-making;
- Removing the need for minor commercial developments that have little wider impact to require planning permission (including commercial microgeneration);
- Supporting the 'town-centre first' policy, but removing the requirement to demonstrate the need for development;
- In the context of the findings of the Lyons Inquiry into Local Government, to consider how fiscal incentives can be better aligned so that local authorities are in a position to share the benefits of local economic growth;
- Ensuring that Secretary of State decisions focus on important, strategic issues, with a reduction by around 50 per cent in the volume of Secretary of State call-ins;
- Ensuring sufficient resources for planning, linked to improved performance, including consulting on raising the £50,000 fee cap and allowing firms to pay for additional resources;
- Enhancing efficiencies in processing applications via greater use of partnership working with the private sector, joint-working with other local authorities to achieve efficiencies of scale and scope, and an expanded role of the central support function ATLAS;
- Speeding up the appeals system, through the introduction of a Planning Mediation Service, better resourcing, and allowing Inspectors to determine the appeal route. From 2008-09 appeals should be completed in 6 months; and
- Improving skills, including through raising the status of the Chief Planner, training for members and officers, and wider use of business process reviews.

Figure 5.4   Key recommendations of the Barker Review of Land Use Planning (© Crown copyright)

Reflective summary

The planning system in England and Wales is based upon decision making at three levels of government, with particular emphasis placed on the district councils. For most forms of development involving householder applications and minor changes of use, the system generally works in a straightforward manner. However, for more major forms of development, the political nature of the system commonly means that the simple exercise of completing an application form correctly is insufficient to obtain planning permission. Consultation between developers and council officers, local interest groups, highway authorities and others on anything but the smallest project is now regarded as critical – and this, combined with the increasing role of the development plan and its adoption process, has changed the way in which developers regard this system. In a general book on property development, the broad principles set out above provide a good basis upon which to approach the planning system. It should be noted, however, that like valuation, tax or building surveying, planning matters are a specialism in which developers often employ expert advice. The heightened concern of government and others in relation to environmental matters is greatly influencing the planning system as well as the EU. This concern has placed town planning in its widest sense in a high place on the political agenda.

6

Construction

6.1 Introduction

The developer's second major financial commitment is to place a contract to construct the development. From this point forward, some of the earlier flexibility will go, although this does, to some extent, depend of the procurement route selected. This is a crucial stage in the development process and the key aim is to construct a good quality building that performs on time and on budget. 'Time, performance and cost' is the mantra. The selection of the procurement route needs to be early on as it has an affect on the composition and the size of the professional team. After the initial brief is decided, a schedule of accommodation is prepared and the broad design constraints are decided, the choice of building contract can be made. This chapter firstly provides an overview of the different types of building contracts and procurement routes, with the second part explaining the management from pre-contract to the post-contract defects liability period. The roles of key personnel such as the project manager are explained in the chapter. In addition, the risks associated with building contracts are identified along with strategies to either reduce risk or to transfer it to the contractor or other parties. The procurement options of public–private partnerships and partnering are also covered in the chapter.

6.2 Procurement

The decision on the form of contract will depend on the developer's requirements and the size and complexity of the development, and, again, time, cost and performance are key influencing factors. The developer must determine whether cost management, time or building performance is the highest priority, as each will favour different procurement options. As you can see from Table 6.1, different types of developer and stakeholder attach varying degrees of importance to time, cost and performance (quality) and these factors will influence the procurement strategy to be selected.

The selection of an appropriate procurement strategy has two elements: analysis and choice. The developer, usually with their consultants, must firstly assess and set the project priorities and, significantly, their attitude to risk. Second, the developer must consider all possible procurement options, evaluate them and choose the most suitable. The factors that have to be taken into account at this stage include:

- factors outside the control of the professional/project team (such as interest rates, inflation and legislation)
- client resources
- project characteristics
- ability to make changes
- risk management
- cost
- time
- performance and quality.

Clearly there may be conflicts between these factors, and priorities need to be set to ensure that the procurement strategy selected gives the client/developer most control over the factors that are of the greatest importance. There are a range of checklists that developers can use to determine which procurement route is most suitable, and readers are directed to the work of Kelly *et al.* (2002) and Morledge *et al.* (2006) for further detailed information.

Table 6.1 Examples of prioritised criteria by client type

	Owner occupier (%)	Speculative developer (%)	Investor (%)
Performance (function/quality)	45	20	50
Time (certainty or speed)	25	50	30
Cost (certainty or price)	30	30	20
Total	100	100	100

Source: Morledge in *Kelly et al.*, 2002

Developers must decide whether the building design is to be carried out by an architect and the professional team, or the contractor. For example, it may be necessary to shorten the pre-contract time by overlapping the design and construction elements of the scheme. Alternatively, early completion may be achieved by using fast-track methods of construction. Of crucial importance in deciding on the form of contract is the likelihood of changes to the design during the contract and the need for flexibility. An additional important factor is the extent to which the developer wishes to pass risk on to the contractor. A public sector developer will be also concerned to achieve value for money to meet the requirement for public accountability.

There are no hard and fast rules about choosing any particular form of building contract. A developer may use any of the available types and adapt it to suit their own particular requirements, provided that it is generally acceptable to building contractors. However, there are advantages in using the forms of contract typically used in the building industry, where there is tested and practical experience, so that the strengths and weaknesses of that particular type of contract are known. From the developer's perspective, building contract arrangements (often referred to as procurement methods, i.e. methods used to both design and construct a scheme) may be broadly divided into three main categories, albeit with many variations in each.

The first is based on the use of the traditional standard form of contract evolved by the Joint Contracts Tribunal (JCT), which provides for a main contractor to carry out the construction in accordance with the designs and specifications prepared by the developer's own professional team – and upon whom they must rely for the quality of design, adequate supervision of construction and suitability of the building for the purpose for which it is designed. This procurement option is also referred to as design–bid–build.

The second category is the 'design and build' contract, which is different and is frequently used in preference to the traditional contract. Here the contractor is responsible for the construction, and also for the design and specification. The contractor takes full responsibility for ensuring that the building meets the requirements of the developer and is fit for the purpose for which it is designed.

Thirdly, there is management contracting, based on American methods of construction, which has been used by some of the larger development companies on complex developments since the 1980s. Like the JCT contract, the professional team is responsible for the design and specification. However, the building work is split into specialised trade contract packages and the management contractor – for a management fee – co-ordinates and supervises the various subcontractors on behalf of the developer.

The traditional (design–bid–build)

Traditionally, this procurement route is based on the contract as drafted by the JCT, which comprises a number of bodies: the Royal Institution of Chartered Surveyors, the Royal Institute of British Architects and the British Property Federation (representing developers and property owners).

Though this contract has been subjected to much criticism, and has been reviewed and amended, it nevertheless continues to be used. It tends to be used on straightforward small to medium-sized schemes. The developer is able to use a number of variations and amendments to tailor the contract to the needs of the project.

The following describes a typical contract.

Developers appoint their own professional team who are responsible for the design of the building to meet their requirements, for supervising the construction phase and for administering the contract. The architect together with the project manager (if appointed) lead the professional team and call in whatever other team members they need to deal with such matters as structural design problems and the provision of mechanical and electrical services. Quantity surveyors or construction economists should be appointed at the outset, not only to provide a cost estimate but to provide cost-planning services. Frequently project managers have a quantity surveying background that enables them to maintain a keen eye on financial management of the development. The role of the professional team in designing and administering the scheme under this type of contract is described below.

The role of the professional team

Unless a planning consultant is appointed, the architect is responsible for obtaining planning permission and all other statutory approvals such as building regulation approvals and fire certificates. The architect is responsible for the design of the buildings in terms of aesthetics and functions, all in accordance with the developer's brief and budget. The architect also is principally responsible for the management of the contract, although supervised by the project manager if one is appointed. The architect does not supervise the building contract on site on a full-time basis. Accordingly, a developer may also appoint a clerk of works or resident engineer at extra cost to carry out a full-time on site supervisory role.

The quantity surveyor is responsible for preparing estimates of building cost, preparing Bills of Quantities (a measured specification of materials and work to enable the contractor to submit a price) and, during construction, for preparing valuations of work (usually monthly) upon which the architect issues the 'interim' and 'final' certificates. The quantity surveyor should be appointed as early as possible within the development process to advise on cost management and the merits of alternative forms of construction (e.g. steel frame or concrete frame) and should also provide estimated cash flows of the building contract expenditure. The quantity surveyor reports on the cost of construction and measures actual payments against the estimated cash flow. Their role is to explain why the actual cash flow differs from the original estimate and prepare revised estimates for the remainder of the project. The quantity surveyor is also responsible for estimating the cost of possible variations in design so that the development team can decide whether or not they are merited.

Other members of the professional team may include structural engineers and mechanical and electrical engineers. The structural engineer works closely with the architect and quantity surveyor, in assisting with the design of the structural elements of the building, calculating loads and stresses, and advising upon how the design of the building should be modified to accommodate them. A mechanical and electrical engineer advises on all the services required such as electricity, gas, water and the design of the heating, lighting and plant installations, and, where air conditioning is to be provided, they will also be responsible for designing the system and liaising with the architect, so that it can be incorporated into the design. Increasingly, they have a key role in reducing the environmental impact of the in-use phase of a building's life cycle through the integration of sustainable building services.

Provided that the contractor executes the building work in a good and workman-like manner in accordance with the architect's drawings and specification in the Bills of Quantities and with any instructions subsequently given to the contractor by the architect, the contractor will not normally have responsibility if the building is not suitable for the purpose for which it was designed. This is irrespective of whether the unsuitability is attributable to faulty design or to some physical inadequacy in the structure. Developers must turn to their architect and other professional consultants for a remedy. On occasions, the respective responsibilities of the professional consultants for an inadequacy or defect in the building are not clear-cut. In such circumstances, developers find themselves in a complex situation, dependent upon the outcome of the arguments between professional consultants. Such situations may be avoided by the exercise of care in the selection of the team of professional consultants and, where a highly complicated or specialised or sophisticated building is involved, it is helpful if the professional consultants have had previous experience of dealing with that particular building type. The quality of the team has a significant influence upon the success of the development project.

Usually, all the professional team, including any project manager, are appointed on a percentage fee basis, either by negotiation or in accordance with their professional body's suggested scale fee. The percentage fee typically relates to the final building contract sum and, from the developer's perspective, this does not provide a financial incentive for the professional team to ensure that the building is constructed on time and within budget, although the professional's reputation is at stake along with their chances of future work with the same developer. A developer may negotiate a fixed-fee basis for appointment. However, the developer must be aware that the professional concerned will invariably include an element in their fee proposal to cover the risk for extra work and, therefore, the developer may not necessarily gain any advantage.

Developers will require the design team (i.e. architect, structural engineer, mechanical and electrical engineer, and sometimes the quantity surveyor) to enter into deeds of collateral warranty for the benefit of investors/purchasers, financiers and tenants. Collateral warranties extend the benefit of the developer's contract with the professionals involved. Typically, they require the professional practice or company to warrant that all reasonable skill, care and attention has been exercised in their professional responsibilities and that they owe a duty of care. In addition, the professionals are required to warrant that they have not specified deleterious materials. They will also need to provide evidence of sufficient professional indemnity insurance (PI) cover from their insurance company. It can often be difficult and time-consuming to procure these warranties in a form acceptable to all parties, as financial institutions and banks require almost total responsibility from the professional design team. The professionals, in turn, are increasingly resisting deeds of collateral warranty in the form required by financial institutions and banks due to the restrictions placed on them by their insurers in relation to their professional indemnity policy. It can become very complex for the developer acting as agent in the middle. For example, professionals may refuse to sign these agreements under seal, because they wish their responsibilities to last for 6 and not 12 years. Additionally, they may wish to limit the assignability of their deeds of collateral warranty to the first purchaser and first tenant of the completed development. They also insist their liability is restricted to remedial costs of any defects, not consequential or economic loss. It is important, at the very least, for the developer to ensure that every member of the design team signs a warranty before they are appointed.

Some developers have turned to insurance in the form of latent defects insurance known as decennial insurance, especially on larger schemes. Latent defects insurance is expensive, typically 1–2 per cent of the building contract sum, but it has the advantage that the insurer is responsible for pursuing remedies with the professional team. The insurer assumes responsibility for repairing the property should an inherent structural defect be discovered that renders the building unstable or threatens imminent collapse. The insurer typically covers a project for up to 10 years from completion, provided that an audit has been carried out before construction begins and an independent engineer reports on the design and construction of the building. The developer has to pay for the fees of the independent engineer. The insurance policy is totally assignable to tenants and purchasers, who seem to be increasingly demanding that such an insurance policy is in place. All residential property development in the UK, unless architect supervised, is covered by either the National House-Builder Council (NHBC) or Zurich Insurance, to protect the mortgagees of latent defects, and the period is usually for 10 years.

The majority of the design work on this type of contract is carried out prior to the appointment of a contractor and has a considerable impact on the final cost. Once the contractor has started work, variations requiring revised instructions to the contractor frequently result in increased costs or delays to the contract programme. Revised instructions are often caused by the developer making a variation, or the architect issuing late instructions as the design is inadequate or incomplete. Thus, the relationship between the developer and the architect is crucial at the design stage. Developers should establish positive and realistic cost limits, which should be relayed to the architect and the professional team. They should ensure that the architect thoroughly understands, with the aid of a written brief, their requirements in respect of all the aspects of the building and its usage, the standard and type of finishes required, the services needed and date for completion. Above all, the developer should avoid, if possible, changing their mind, unless this is necessary to secure a particular tenant or improve the value of the scheme. It is vital that the architect is chosen with great care as much depends on the architect producing the working drawings for the contractor on time, and the developer must determine whether an architect has the capability and resources to deliver the design element effectively.

Discussion points

- What are the key professionals developers need to bring into the development team?
- What roles do they play in the team?

Choosing the contractor

Once the detailed design is completed, the quantity surveyor prepares the Bill of Quantities, which specifies and quantifies the materials and the work to be carried out in great detail – right down to the number and make of door locks. All Bills of Quantities must be checked as any errors can involve additional cost due to later variations to the contract. In addition, provisional sums and contingency sums must be included. After this, contractors are invited to submit tenders for carrying out the work based on the drawings and the Bill of Quantities. However, this is not necessary: there is nothing to prevent a contractor being asked to price the Bill of Quantities. This may happen when the developer has employed a particular contractor (it may be an in-house contractor) over time and is pleased with the quality of their work and, therefore, prefers to re-employ them. It may be that the contract contains a great deal of specialised work for which one contractor has an outstanding reputation and they might be chosen on that ground to carry out the work, subject to a acceptable pricing of the Bill of Quantities. If this route is taken it may be advantageous to appoint the contractor earlier in the development process to advise the design team on the practical aspects of the design.

When choosing contractors to be invited to submit competitive tenders for carrying out the work, there are issues to be considered. It is necessary to limit the total number of contractors invited to submit competitive tenders. Six or so contractors are often adequate for even the largest contract; this is because the pricing of Bills of Quantities for a large job is time-consuming and expensive for contractors who are not keen to submit competitive tenders when there is, in their opinion, an unreasonably large number of tenderers.

If the work is specialised, contractors skilled in that type of work are selected. Sometimes it will be preferable to use a large national contractor, while at other times local or regional contractors are favoured. Some contractors have a reputation for producing work of high quality, others for producing work quickly and on time, and this can be a vital consideration for the developer's cash flow. Still other contractors have a reputation for submitting keen tender prices, and there are those who have a reputation for expertise in formulating claims for extra payments on any and every occasion during the contract. The architect or developer may feel that some contractors are entirely dependable, while others may have let them down on a job in the past. Unfortunately, high-quality work, speed and low cost are a very rare combination. Developers are often guided by their architect, project manager and/or quantity surveyor on the selection of the contractors for the tender. Before a contractor is included, they should be asked whether they are willing to tender for the job. There are times when contractors are fully extended and are unwilling to tender for work. Some contractors, for a variety of reasons, may not be interested in tendering for work in a particular locality.

The prices submitted by each of the contractors are examined by the quantity surveyor and the project manager to ascertain what is offered. The quantity surveyor prices the Bill of Quantities independently and this is used for comparison purposes with the contractor's tenders. Each contractor prices each element of the work, setting out the applicable rates for each element of work or unit of material. The contractor's priced Bill of Quantities forms part of the contract documentation and is used by the quantity surveyor to value the work carried out by the contractor.

The reliability and financial stability of the contractor are vital considerations. Therefore, when contractors are chosen, it might be advisable to take out a 'performance bond'. It is the contractor who takes out the performance bond with an insurance company who guarantees to reimburse the employer for any loss incurred up to an agreed amount as a result of the contractor failing to complete the contract. Financiers may also insist on a performance bond for their benefit. The failure of a contractor is a major disaster from the developer's perspective as long delays occur while the legal position is resolved and another contractor found to complete the work. The new contractor might ask a considerably higher price for completing the job than was contained in the original contract. Furthermore, if defects later appear in the completed building, it can be very hard to apportion responsibility between the original contractor and the contractor who takes over the job. Thus, it is easy to understand why employers ask for a performance bond, even though the extent to which their losses are reimbursed is limited and the cost of the performance bond is usually added to the contract cost.

Paying the contractor

The method of payment for the works has a substantial impact on the developer's cash flow position. It also has significant impact on the contractor's cash flow, who has had to consider the method of payment when preparing the price for the work. Thus, the method of payment must be made clear when the contractor is invited to tender.

Under the JCT contract, the architect usually authorises monthly payments based on the value of work certified by the quantity surveyor. Usually a certain percentage (3–5 per cent) of the total value of the work undertaken is retained until the end of the contract. This is known as the retention. This arrangement best suits the contractor who obtains payments for the work carried out irrespective of when the building is ready for occupation. The developer has to pay out very substantial sums of money over a considerable period of time before obtaining the benefit of a completed building at the end of the contract.

The ideal arrangement for developers is for the whole of the contract price to be paid when the building is handed over, so that they do not part with their money until the time when they should be receiving an income from the building or have the benefit of occupation of it. It must be remembered, however, that if it were possible to make such an arrangement (it is extremely rare), contractors would increase their tender prices by one means or another to take account of interest and the additional risk. In the case of a large contract spread over a period of time, some contractors might not be able to finance the work easily without payments from the developer. Some compromise might well be devised for payment to be made in certain set stages – the last payment on completion and handover of the building is weighted to give the contractor an incentive to get the building completed. The method of paying for the work has to be related to the circumstances of each contract. The contractor is more likely to be flexible if they are a partner in the scheme and stand to benefit in terms of a profit share.

Calculating the cost

When using the JCT form of contract, the contractor usually submits a bid on either a 'firm price' or a 'fluctuations' basis. The firm price means that although the cost of labour and materials used in carrying out the work may fluctuate with the market, the contract sum will not be varied to take account of these fluctuations. The fluctuations basis means that once the contract is awarded to the contractor any increase or decrease in labour and materials is added to or subtracted from the contract sum. Under both types of contract, there is a clause allowing adjustments to be made to take account of alterations in cost due to government legislation. However, the developer may delete such clauses, particularly when the tender market is very competitive. It is vital to note that firm price does not mean the contract sum once fixed will not alter. Quantity surveyor's remeasurements, architect's variation orders and instructions and extensions of time may affect the cost.

Developers and their professional advisers must decide on what basis they wish contractors to prepare their competitive bids in order that the contractors submit prices on the same basis. Developers do not always find it easy to decide which basis is likely to be to their advantage. The risk of fluctuations in the cost of labour and materials during the contract cannot be avoided; the question to be decided is whether the risk is to be borne entirely by the contractor as in a firm price contract, or whether the risk is to be borne by the developer as in a fluctuations contract. If contractors have to prepare their bids on a fixed-price basis, they will add something to their prices to cover themselves against the risk of increased costs. To tackle this problem, contractors are asked often to quote prices on both a fixed and fluctuations basis, leaving the developer to decide in the light of the differing prices which basis is likely to prove most advantageous during the whole of the contract period. Each contract must be judged on its own merits against the background of tender market conditions.

There are two main alternative methods of calculating the cost. Firstly, a cost-plus contract, where the contractor is paid on the basis of the actual cost of the building work ('prime cost') plus a fee to cover their overheads and profit. The fee might be fixed or a percentage fee calculated with reference to the final building contract sum or the initial estimate. Secondly, a target cost contract might be negotiated or established by tender. A target cost is agreed with the contractor plus the contractor's fee. Any savings or additions to the target cost are shared by the parties.

The duration of the contract

The date agreed in the contract for the completion of the building is not certain as the contractor can apply for extensions of time for a number of reasons. Some of the reasons that justify an extension of time entitle contractors to recover additional loss or expense that they may have suffered as a result of that extension. Extensions of time usually result in an increase in the cost, which will certainly include the contractor's 'preliminaries' (overheads such as insurance, cost of plant hire etc.) The impact of an extension is felt twice by the developer: firstly, it affects cash flow and, secondly, it increases costs.

The main reasons for extension of time that entitle the contractor to recover additional loss and expense are:

- Inadequacy in the contract documents: the drawings and/or the Bill of Quantities. This may be due to professional incompetence in the preparation of the documents or new legislation might be introduced during the contract that requires amendment to the drawings. Additionally, the building control officer of the local authority (responsible for providing building regulation approval) or the local fire officer (responsible for issuing a fire certificate) might impose conditions on their approval that necessitates design changes. Unforeseen ground conditions (in the case of a cleared site), hidden structural problems (in the case of a refurbishment) or contamination issues can cause additional expense and delay. This underlies the necessity for a thorough site investigation and/or structural survey before the tender documents are prepared.

- Delay by the architect in issuing drawings or instructions.
- Delays caused by tradesmen directly employed by the developer.

Additional reasons that may be included in the contract terms (depending on the results of negotiation with the contractor on the standard clauses), entitling the contractor to an extension of time, but not to recover any additional loss or expense, include:

- Failure by the nominated subcontractors: on almost every building contract some work is subcontracted to specialists. When contractors find and appoint their own subcontractors, they are responsible for delays caused by them, so that the developer does not suffer. Architects can nominate subcontractors to undertake certain elements of work. The reasons for doing this may be the high quality of previous work with the architect, or expertise in designing and constructing elements of the work, e.g. the structural steel work.
- Bad weather: architects should ensure an accurate record is taken of weather conditions on site.
- Strikes and lockouts.
- Shortage of labour (tradesmen) or materials.
- Damage by fire where it is the contractor's responsibility for insurance under the contract.
- *Force majeure*, i.e. acts of God such as earthquakes, floods, storms etc.

Developers may be compensated for a delay for which the architect has not granted an extension of time under the terms of the contract. The compensation is in the form of liquidated and ascertained damages at a rate agreed in the contract to cover the developer's loss. However, the agreed rate can fall short of the developer's true loss in terms of the overall development cash flow.

A developer cannot assume that the work will be carried out within the time set out in the contract and for the exact contract sum. A key advantage of JCT contracts is flexibility with regard to the way the price for the job is to be fixed and its elasticity, which enables the type and quantity of work within the contract to be varied and yet leave the quantity surveyor free to negotiate the final price for the job at the end of the day. However, there is a disadvantage with regard to the flexibility of this contract: it does not discipline the developer into making early clear-cut decisions as there is always the scope to make late variations. Also the competence and efficiency of the professional team is vital to the successful outcome of the contract. Inadequacies of the plans and specifications are not realised until the building contract is underway. Also, the professional team is not motivated sufficiently to control costs and delays, as their fees increase in proportion to the final contract sum. This type of contract may lead to a very confrontational situation with the contractor if the tender documents are inadequate or many variations are made. Also, many contractors have a tendency to use the claims procedure as a negotiating ploy to claim additional money to cover losses they have made on the contract. It is for this reason alone that many developers use design and build contracts. Developers must ensure that they are in control and are kept closely advised as to the likely financial outcome of the contract at all times. The best arrangements for dealing with this are examined in 'Project management' below.

Overall the key advantages and disadvantages of the traditional approach can be summarised briefly as shown in Table 6.2.

Discussion point

For what reasons might a developer select a traditional (design–bid–build) contract for a development project?

Table 6.2   Advantages and disadvantages of traditional (design–bid–build) procurement

Main advantages	Main disadvantages
Competitive fairness	Strategy is open to abuse resulting in less certainty
Design led – enabling high quality design	Overall programme may be longer than other options because there is no parallel working
Reasonable price certainty based on market forces	No 'buildability' input from contractor
Acceptable strategy in terms of public accountability	The strategy can lead to adversarial relationships between the parties
Well known procedure	
Flexibility - changes re easy to arrange and value	

Design and build

This is radically different: one party, typically the contractor assumes the risk and responsibility for the design and construction of the scheme, in return for a fixed-price lump sum. Design and build is a fast-track strategy. Its use has become more widespread due to dissatisfaction with the traditional JCT contract, and the problems encountered with splitting design and construction responsibilities. Design and build was used on simple and straightforward schemes originally but is now used on most types of building. It is widely used by public sector clients for hospitals and schools etc. The standard contract used for design and build is the 'JCT 1998 WCD (With Contractors Design)'.

The contract is based upon a performance specification by or on behalf of the developer. Here 'performance' means the various requirements that the building must meet. However, the developer's requirements must clearly, and in as much detail as possible, be set out in the performance specification if this type of contract is going to work from the developer's viewpoint. The responsibility for the design or what precautions should be taken to ensure that the finished building meets with all the various statutory requirements and suitability for the purpose for which it is designed rests wholly with the contractor.

Performance specifications vary from being fairly simple to very detailed depending on the nature of the scheme. For example, a developer may wish to develop a site with very simple standard-design warehouse units. The specification in this case may include a schedule of floor spaces for units of different size, with an indication as to how much office accommodation is to be provided, the total amount of toilet accommodation, the services to be put into the building, the floor loadings and the clear floor heights, together with an indication of the total yard area. On that simple performance specification, contractors would be asked to submit schemes for the erection of their own standard-design units to meet the requirements, together with their price for carrying out the contract. The complete responsibility for obtaining all the necessary statutory approvals, and for designing and erecting the buildings and ensuring that they will be suitable for the purpose for which they are required, rests with the contractor.

However, the performance specification tends to be extremely detailed and will specify the materials to be used by the contractor. It is the contractor's responsibility to comment on any materials specified by the developer if they consider them to be unsuitable for the purpose, before the contract documentation is entered into.

It is possible to arrange for complicated buildings to be erected under a design and build contract. However, in such cases, the performance specification is critical and has to be carefully prepared by a team of professional advisers. In the case of a complicated building, which does not conform to a standard design, developers are entirely dependent upon the adequacy of their own performance specification to ensure that they get a building that meets their requirements.

A developer using this type of contract may appoint a specific contractor with which they have successfully worked before or who has expertise in constructing buildings similar to the one proposed. Alternatively, a developer will go out to tender to appoint the most suitable contractor for the job based on their design, specification and price. Typically, the tender list will be short (two or three) due to the considerable amount of work involved by the contractor. As with the traditional JCT contract, the track record and financial stability of the contractor is an important factor. Some contractors specialise in design and build, building up valuable experience in constructing and designing, while some larger contractors have 'design and build' divisions. The contractor may employ all the necessary skills in-house or, more usually, employ external architects and engineers under the supervision of their own in-house project managers.

A developer may appoint a quantity surveyor to carry out the usual pre-contract activities such as advising on the most suitable form of contract and preparing initial cost estimates. The quantity surveyor will perform similar duties during the contract as those on a traditional JCT contract, including valuations for interim payments and the final account.

In some cases, the quantity surveyor may take on the role as the developer's representative (referred to as the 'employer's agent' in the contract), and effectively project manage and administer the contract. The quantity surveyor may agree the letters of appointment and deeds of collateral warranty with the professional team, and chair and minute all project meetings (usually the role of the architect under a traditional JCT contract).

Under a design and build contract the contractor submits drawings and specifications to the developer for approval, who can check to see just what type of building will be built, what services will be provided, and so on. The contractor is responsible for designing and constructing the scheme in accordance with the approved drawings and specification. In a simple package deal where a standard type of building has already been erected by the contractor in a number of locations, and examples of which can be inspected and the occupiers asked for their comments on the adequacy of the building, developers will know the product and any feedback on its suitability. In such circumstances, the design and build contract is advantageous where the buildings are of a simple, straightforward nature, with often repetitive design elements, and can be built to a standard design that has been used by the contractor elsewhere. The advantages are that the design time and cost can be greatly reduced: the contractor is working to their own designs with which they are familiar; and they may use various standard types of component, which they can buy advantageously and which they are used to using on site. The contractor's own designs will undoubtedly reflect the contractor's practical experience of putting up buildings. The result should be that the contractor works more efficiently and speedily, and thus more economically, so that the price of the building to the developer ought to be lower.

The advantages of this type of contract for the developer are that while it is possible to provide for fluctuations in the contract price, and there are various alternative ways of paying for the buildings as the contract proceeds, usually a lump-sum fixed price is agreed: the contractor is committed to provide the buildings for a known cost and takes the risk. Clearly, contractors allow for these risks when preparing their price but the developer is reassured to know that the price is fixed. It may be changed only by variations issued by the developer or changes in legislation. When the tender market is competitive clauses in the contract dealing with changes in legislation may be deleted. The developer does not run the risk of becoming involved in endless professional arguments if the design or construction of the building is defective; it will be entirely the responsibility of the contractor to see that matters are remedied.

There are disadvantages, for example the developer does not have the same detailed control over the design, and if the developer requests alterations during construction the cost might be increased out of all proportion – in short, the developer does not have the protection of the flexibility of the traditional JCT contract. It can be argued that the final cost of the buildings under a fixed-price, lump-sum package deal may be higher due to the risk that is carried by the contractor but, in practice, this is often offset by the advantages to the contractor of using his own design and standard components. Another advantage is the likely achievement of an overall saving in time due to the overlapping of the design and construction processes. Furthermore, the developer will save money on professional fees as involvement by professionals will be less.

There may be types of development for which the design and build type of contract is not suitable. Some developers believe that the aesthetics and quality of the finished building is lower compared to the traditional JCT contract. To overcome this problem a developer can appoint an architect to prepare the initial drawings and sketch designs under what is known as a develop and construct contract. The contractor is then responsible for developing the design as part of their tender submission. However, under this arrangement the developer must ensure that the design responsibility is adequately defined.

Alternatively, the developer may appoint an architect and engineer and novate their appointment contracts to the contractor. By novating the contract the contractor steps into the shoes of the developer and becomes the client of the professional concerned under the terms contained in the original contract. Thereafter, the contractor takes control of the design process and the professional is liable to the contractor. A potential conflict of interest may arise as the architect/engineer may continue to treat the developer as their 'employer' on the basis of their long-standing relationship. It is important to ensure that the professional and the contractor develop a good working relationship. The developer will still insist on deeds of collateral warranty with each professional in case the contractor goes into receivership. The advantage to the developer under this variation is that they can appoint their choice of architect and engineer, with whom they have a good working relationship, whilst retaining the advantages of the standard design and build contract. A key drawback of the design and build contract for developers (whichever variation is adopted) is the lack of flexibility. Developers must decide before the contract is signed on their exact requirements, as major variations can usually only be made at a considerable cost. In summary, the key advantages and disadvantages of the design and build approach are briefly as shown in Table 6.3.

Management contracting

With this procurement strategy, a management contractor is engaged by the developer to manage the building process and is paid a fee. Management contracting, developed in the USA, became more widespread in the UK during the 1980s because developers were impressed with the fast-track methods of construction. The management contract is generally used on larger, complex development projects where the developer requires speedy construction at competitive prices with the flexibility to change the design during the contract. In summary, the building contract is split into specialised contract packages, either by trade or building element, and let separately under the supervision of the management contractor.

The developer appoints the professional team in the usual way (as described in 'The role of the professional team') to prepare the drawings and specification for the project. The quantity surveyor prepares a cost plan based on the drawings and specification. The actual cost incurred by

Table 6.3   Advantages and disadvantages of design and build procurement

Main advantages	Main disadvantages
Developer deals with one firm	Fewer firms offer design and build so
Inherent buildability is achieved	there is less real competition
Price certainty is achieved before construction commences (provided the developers requirements are adequately specified and changes are not made)	Developer has to commit before detailed design is completed
Reduced total time of project due to overlapping activities	In-house design and build forms are an entity, so compensation for weak parts of the firm is not possible
	There is no design overview unless separate consultants are appointed for the purpose by the developer
	Preparing an adequate brief can be difficult
	Difficulties comparing bids as each design, programme and cost varies
	Design liability is limited by the standard contract
	Changes to project scope can be expensive

the management contractor ('prime cost') is paid by the developer, having been certified by the architect and monitored by the quantity surveyor. The developer pays a fee to the management contractors for their services in managing the various separate contracts. The developer may not always appoint the contractor with the lowest fee proposal. It is crucial that the contractor has management contracting experience and sufficient staff with the right skills. The fee may be a lump sum or a percentage of the contract cost plan. The prime cost includes the amounts due to the various contractors of the various parts of the project, plus the management contractor's own on site costs. The construction work is carried out by the various contractors, who enter into a standard JCT contract with the management contractor based on detailed drawings, specifications and Bills of Quantities. The selection of the contractors should be carried out by the developer and the professional team in consultation with the management contractor. The architect has the power to issue variations known as 'project changes'.

The management contractor should be involved at an early stage with the professional team in advising on the practical implications of proposed drawings and specifications, and the breakdown of the project into the various separate packages. The management contractor is paid in relation to interim certificates issued by the architect, including instalments of the management fee. The disadvantages are that the developer has to pay the management contractor's fee, as well as the professional team, and the management contractor is not responsible for the actual building works. The developer has no direct contractual relationship with the various contractors carrying out the work. Accordingly, the developer must enter into design warranties with each contractor that are capable of being passed to purchasers, financiers and tenants. However, the management contractor may have to pursue the remedies of the developer in respect of any breaches by the various contractors. The developer has to reimburse the management contractor in settling or defending any claims from the contractors, unless the management contractor is in breach of the contract or of their duty of care. The liabilities and hence risk of the management contractor is limited. They are not responsible for the payment of any liquidated damages for any cost overruns if caused by reasons outside their control or by delays due to the various contractors. It is essential that the management contractor, developer and the professional team co-operate since the developer has to pay extra for the management contractor's expertise and experience.

Note that the management contract itself is not a lump-sum contract. The management contract is based on the contract cost plan prepared by the quantity surveyor, which is only an indication of the price. However, the contracts the management contractor enters into with the various contractors are usually based on the standard lump-sum JCT contract. Therefore, the final cost is based on the contracts with each specialised trade and it may bear no relation to the cost plan within the developer's contract with the management contractor. Cost control is essential under this contract and it must be very tightly managed by the project manager and quantity surveyor as there is no direct incentive for the management contractor to keep within the cost plan as their fee is directly related to the final building cost. The success of this type of contract in terms of cost control depends on the ability of the management contractor to appoint the various trades within the budget of the cost plan. However, the final cost is whatever the cost is to the management contractor (including the fee) and there is no penalty for exceeding the cost plan.

The developer has to accept a very high degree of risk with this type of contract. Developers who have used this type of contract have found cost control the biggest problem as there is no tender sum. There is no control over the delays caused by the individual contractors. There are extra costs involved in duplicating site facilities for the management contractor and the various contractors. However, the main advantage of management contracting is speed as projects are usually completed more quickly than on traditional contracts where full detailed drawings have to be prepared before the contract commences. This speed is achieved by the flexibility of dividing up the contract into separate elements, overlapping the design and construction of each element. The 'packaging' of the contract allows the developer, to some degree, to control costs and delays as contracts let later on in the process can be varied to suit. Pre-construction and construction times can be reduced when compared to other contracts. Developers are concerned that this type of contract does not effectively control the development's design and quality standards suffer as a result.

Due to some negative experiences with management contracting due to the disadvantages noted above, a variation known as 'construction management' was preferred towards the end of the 1980s. With construction management, trade contractors are placed directly by the developer and a construction manager is appointed for a fee as part of the professional team appointed at the same time, not necessarily afterwards like the management contractor. The most famous office development of the 1980s, Canary Wharf in London Docklands, used construction management.

The construction manager acts as the developer's agent and the appointment of a project manager is required to co-ordinate the professional team. Their fee is usually percentage-based with an additional lump sum for the provision of site facilities. Their role is to manage and co-ordinate all the various contractors, review design proposals, control costs (against an agreed budget), control the contract programme and be responsible for quality control. The administration of claims for payment by the contractors and variations are their responsibility, although the final account with all the contractors is administered by the architect and quantity surveyor.

The advantage of employing a contractor on the professional team is to bring their experience and expertise into the design stage at the beginning. A contractor may employ 'value engineering' techniques (detailed studies of the cost-effectiveness of alternative materials and methods of construction) to review the design process. However, the developer has to ensure that the contractor has the relevant design experience, otherwise the advantage of strict design control will be negligible. The developer has to have much greater involvement in this type of contract arrangement and it is the job of the construction manager to ensure that the developer makes firm decisions at the appropriate time. It is a very management intensive contract resulting in higher staff and fee expenditure on behalf of the developer compared to other types of contract.

The key advantage is the saving of time, achieved by overlapping the design and construction of each package and involving the construction manager at the beginning of the design process. It is appropriate where an early completion of the scheme is crucial. However, despite the developer having direct control over the various contractors, cost control remains a problem. There is still no guarantee of what the final cost of the scheme will be, although incentives may be used to increase the contractors' share of the risk but at a cost to the developer. This method may be used by developers if they wish to maintain flexibility, take advantage of fast-track methods of construction and retain control while accepting greater risk. They could reduce this risk if they were confident of their exact requirements at an early stage and the pre-contract period was long enough to allow for detailed design.

Overall the key advantages and disadvantages of the management contracting approach can be summarised briefly as shown in Table 6.4.

A similar strategy to management contracting is the design and manage contract. Here a contractor is paid a fee to manage and assume responsibility for the works and also the design team. The advantages are early completion because of overlapping activities; the developer deals with one firm only; it can be applied to complex buildings; and the contractor assumes the risk and responsibility for integration of the design and construction. On the other hand, the disadvantages of this approach are: price uncertainty (not achieved until the final work package is let); the developer loses control over design quality; and the developer has no

Table 6.4 Advantages and disadvantages of management contracting procurement

Main advantages	Main disadvantages
Time saving potential for overall project time	Need for a good brief
Buildability potential	Uncertainty on price
Breaks down traditional adversarial barriers	Relies on a good quality team
Parallel working is inherent	
Work packages are let competitively	
Flexibility – changes can be made provided the packages affected have not been let and there is little impact on those already let	

direct contractual relationships with the work's contractors or the design team – thus making it difficult for the developer to recover costs if they fail to meet their obligations.

6.3 Project management

The appointment of a project manager is not necessary for every project. A project manager tends to be needed for large and complicated rather than small, simple projects. Often, developers act as their own project manager with 'in-house' staff or employ one of the professional consultants to exercise the management function. Typically, a project manager will receive a fee representing 2–3 per cent of the final building cost, depending on the extent of the role and the complexity of the scheme. However, the developer may appoint a project manager on an incentive basis linked to whether the final cost is within budget. A development company may be asked to take on the role of project management on the basis of a fee, either fixed or related to the profit of the development, by an owner–occupier or property investment company for instance. Project management in this context has a much wider definition and includes the management of the entire development process.

Project management is an occupation and project manager's may be architects, quantity surveyors, valuers/agents or have a building/contracting background. The project manager should be appointed at an early stage to be able to advise the developer with the quantity surveyor on the type of building contract applicable to the development, and to be involved in the development brief and the design discussions. Also, the project manager should be able to advise the developer on the selection of the professional team, particularly those who have previously worked with the project manager. The professional team should complement each other and work well together. The project manager's role is to act as the client's representative when co-ordinating the professional team and liaising with the contractor. The project manager is concerned with the overall management of the project and is not involved in carrying out any part of the project. The project manager needs plenty of common sense, administrative ability and a good knowledge of construction.

The management objectives must be clearly defined in consultation with the project manager and made known to everyone in the project team. The objectives are to ensure that the finished project is suitable for its intended purpose, that it is built to satisfactory standards, that completion occurs on time and that the project is carried out within the budget. The project manager is often responsible for appointing the professional team on behalf of the developer and will agree the fees, letters of appointment and deeds of collateral warranty under the guidance of the developer. The developer should ensure that the project manager is supplied with copies of all the funding documentation entered into with any financier of the scheme. The documentation will include the plans and specifications agreed with the financier, and the project manager should ensure that these are complied with throughout. If alterations are necessary, then the approval from the financier will be formally required. The project manager is responsible for ensuring that arrangements for the disposal of the building, either the letting or the sale, are carried out efficiently and satisfactorily.

Discussion points

* What are the benefits of having a project manager in the development team?
* In what circumstances might a developer decide to appoint a project manager?

It is essential to examine the role of the project manager through the pre-contract and contract stages. The following is based on a traditional JCT contract.

Pre-contract preparations

The project manager should check that the developer has the necessary legal title on the site, whether it is freehold or leasehold, and that vacant possession of the whole site is available immediately. All restrictions on the site should be carefully checked (e.g. underground services, easements and rights of light or support), and compared with the proposed scheme so that the building work will in no way interfere with them. The project manager will arrange (if not previously carried out by the developer) all the necessary ground investigations, structural surveys and site surveys, and communicate the results to the rest of the professional team. It is important that all the site boundaries are clearly defined, and that a schedule of condition of the boundary fences, adjoining roads and footpaths etc. is prepared. It may be necessary to negotiate 'rights of light' or party wall agreements with adjoining landowners/occupiers.

The architect is responsible for ensuring that all the necessary statutory approvals have been obtained, such as planning permission and building regulations. The fire officer should be consulted early in the design process and the architect should ensure that the design is in accordance with all relevant legislation. The architect is responsible for assuring the project manager that all necessary statutory consents have been obtained. It is most important for the project manager to obtain unqualified assurances on these matters because, in practice, many expensive delays are caused as a result of one or other of the statutory consents not being obtained before the contract starts. Sometimes there are circumstances that might persuade the project manager to allow a contract to start before all the statutory consents have been obtained, but in so doing the project manager and the developer must realise the risk that is being taken.

Preparing the contract documents

The project manager's most important job is to ensure that the contract is not allowed to commence without adequate documentation. Incomplete drawings are probably the most common cause of delays and cost increases. If a contract is started before all the drawings are completed and the architect is unable to provide all the drawings to meet the contractor's required time schedules, the consequences can be serious. The project manager needs to be absolutely satisfied with the availability of the drawings by the architect and that sufficient staff resources within the architect's firm are in place. If a contract is started before the drawings are fully complete, which is often the case, a detailed schedule must be obtained from the architect, showing exactly when the outstanding drawings will be delivered to the contractor. Before the building contract is placed, the architect must obtain from the contractor a written statement confirming that (provided the drawings are supplied in accordance with the architect's schedule) there will be no claims for delays due to lack of drawings; getting this matter right at the outset cannot be overemphasised.

Project managers must also be satisfied that the Bill of Quantities is as complete and accurate as possible. The quantity surveyor will measure the quantities off the architect's drawings, so again this stresses the need for their accuracy. Some items in the Bill of Quantities may be described under the headings of prime cost (meaning actual cost) or 'provisional sums'. Prime cost items usually cover materials or goods that generally cannot be precisely defined. Provisional sums items cover elements of the work that it is not possible to detail properly and evaluate at the time the contract is entered into. The contractor is required to allocate a sum of money against these items. The project manager must understand why the prime cost and provisional sums items have been included in the Bill of Quantities, and be satisfied that it is impossible to make the detailed provision at the outset. Quantity surveyors should be questioned to ensure that they have received adequate information from the architect to enable them to prepare their Bills with complete confidence in their accuracy.

If a pre-letting has been achieved then it is important to include any specific requirements of the tenant within the contract document. Furthermore, such requirements should be clearly referred to as the 'tenant's specification' so there is no doubt. There may be a situation where the tenant subsequently alters their specification, which delays the main contract on the 'developer's specification'. If a claim is subsequently made by the contractor then it can be apportioned to the tenant for payment.

Appointing the contractor

If it is proposed to invite competitive bids from selected contractors, the project manager should agree with the architect and the developer the names of the contractors who will be invited to tender. When the competitive tenders have been received and evaluated, the job is normally awarded to the lowest tenderer. However, there has been a move away from automatically accepting the lowest tender with the advent of 'best value' approaches, whereby other factors are considered and the lowest tender may not offer value for money in the long term. Some clients appreciate that paying more in the short term has much greater long-term benefits. The quantity surveyor compares each tender against their priced Bill of Quantities. The project manager decides whether a performance guarantee bond has to be obtained by the contractor. Once satisfied on all matters, the project manager then authorises the placing of the building contract. All the contract documents should be ready, so that the contract may be signed before work actually starts on site, although in practice work often starts before the documents are signed, on the basis of a letter of intent, but this should be avoided. The project manager will discuss with the architect reasons for wishing to appoint any nominated subcontractors and, if appropriate, then authorise their appointment.

Site supervision

The project manager should be continually satisfied about the arrangements made by the architect for site supervision during construction. The size and complexity of the scheme may merit the appointment of a full-time site supervisor, such as a clerk of works or a resident engineer or, indeed, a resident architect. The architect should also arrange for progress photographs to be taken periodically on site, so that a clear visual record of the state of the contract at any time is always available to supplement the architect's own reports on the progress generally.

Construction period

When the contractor has taken possession of the site, the project manager ensures that the works are carried out on schedule and that the overall cost is kept within the budget. To carry out his duties effectively, regular meetings of the project team are held. The frequency and composition of the meetings depend upon the size of the particular job and may vary at different stages of the job. The project management meetings are often arranged on a monthly or fortnightly basis. The project manager, the architect and the quantity surveyor form the nucleus of the project management team. If the project manager is also controlling the letting or sale of the project, then the surveyor/valuer/agent is normally a member of the team, particularly in those cases where the purchasers or tenants might wish to have special works carried out. If the scheme is being financed externally, then the fund or bank's representative or appointed advisers will also attend the project meeting to fulfil their monitoring role. The contractor may be invited to attend the part of the project management meeting at which the progress on site is discussed. Often, the project manager attends separate site meetings with the architect to be kept informed of building progress. All project management meetings must be accurately minuted.

Typically, at the beginning of a project meeting the minutes of the previous meeting are considered and any matters arising dealt with. The architect presents a report on the progress of the work, indicating what parts of the work are ahead of, or behind, schedule and comments on the overall progress of the job. The architect should state any difficulties that have arisen at every meeting and whether the contractor is delayed as a result of lack of information. The architect should also report as to whether any variation orders or instructions have been given to the contractor and, if so, their likely affect on the progress of the work. The project manager will learn independently from the contractor or through attendance at site meetings whether the contractor is being delayed by lack of information or materials/labour.

The quantity surveyor then presents a report on the financial situation, indicating whether or not the work of measurement on site is well up to building progress and whether any variation order or architect's instructions have been given that affect the cost of the job. The quantity surveyor should indicate the position with regard to prime cost and provisional sums and present an overall summary as to how the cost of the job so far compares with the contract sum. The quantity surveyor should also indicate any factors that might increase or decrease the cost of a job at a future date.

When appropriate, the surveyor reports on the progress with regard to the disposal of the property and on any requests for special or extra work received from prospective purchasers or tenants. Then the practicability and advisability of carrying out those special works are discussed. Ideally, purchasers or tenants should take over the completed building in accordance with the original design and specification, carrying out required special works at their own expense once the building has been handed over to them. However, it is not always possible to insist on such an arrangement, as it may be necessary to carry out such works in order to secure the letting or sale.

Then the project manager summarises the overall financial situation, particularly with regard to payments to the contractor, compares them with the budget, checks on dates of handover and compares the estimated date for the receipt of income or capital payments with the budgetary expectation. These are matters of vital importance to the developer's cash flow. If it appears that the project is running behind schedule, then methods of speeding up the work to recover the position are considered, together with the implications for cost. Usually, there is a liquidated damages provision in the building contract and the question of its enforcement has to be considered. In practice, liquidated damages are often inadequate to compensate the developer for losses incurred as a result of the delays, because if the true cost is written into the contract documents at the time of the invitation of tenders, contractors would increase their tender prices out of all proportion in order to safeguard themselves against a the risk of a heavy liquidated damages claim, which might in fact never be made.

This summary of project management arrangements where a JCT contract is used is intended to illustrate the basic principles involved, which also apply generally to other procurement methods. The project team may be much larger and a management contractor or a construction manager may be part of the team on building schemes of a complicated nature. Accordingly, the project management is that much more intricate. On the other hand, where design and build contract methods are used, the role of project management is simplified, and may be taken over by the quantity surveyor. In the case of a lump-sum, fixed-price design and build contract, project managers are essentially concerned with quality control and progress. They may inspect the buildings during construction or arrange for a professional adviser to do so. Periodic meetings with the contractor to discuss building progress and the achievement of the handover dates should enable them to fulfil their role.

Handover

A short time before the date for completion and handover of the building from the contractor to the developer, the architect prepares a 'snagging' list, indicating all the minor defects that must be remedied before handover occurs. It is useful for the developer's surveyor and the intending occupier's representative (if known) to accompany the architect to ensure that all are satisfied with the snagging list. At the outset of the contract, the project manager will have confirmed that the building works are adequately protected by the contractor's own insurance arrangements. The contractor's insurance no longer protects the building once it has been handed over, so it is vital for the project manager to ensure that the developer has adequate insurance cover from handover until the insurance cover provided by the occupier takes effect.

If the development has been pre-let or pre-sold to an owner–occupier, then the occupiers and/or their contractors may wish to have access before formal handover by the main contractor working for the developer. From the developer's perspective, this situation should be avoided, unless it is necessary to secure the deal. If an occupier wishes to gain early access to attend to fitting-out works, then arrangements should be documented clearly in the contract. Ideally an occupier's special requirements should be incorporated at an early stage into the design process or if the occupier is secured after the building contract has started, then the occupier should be allowed access only after practical completion of the building. It is important to include the developer's base specification for the scheme into any documentation, so that any changes that lead to an increase in cost or delays can be attributed to one party or the other. If there is an overlap between the main contractor and the fitting-out contractor, the project manager should ensure that the occupier arranges adequate insurance. Problems can occur when the fit-out contractor's work affects the work of the main contractor. The project manager, with the architect, needs to attribute and resolve problems quickly.

The quantity surveyor should then be asked when any outstanding remeasurement work will be completed and be in a position to agree the final account with the contractor, so that the architect may issue a final certificate. The JCT contract will have provided for a certain percentage of the total cost to be retained by the building owner until the end of the defects liability period, often 6 months from the date of practical completion (12 months in the case of any electrical and mechanical element of the contract). Special maintenance periods may be agreed for particular parts of the work (e.g. landscaping). The contractor is responsible for remedying any defects (other than design) that have occurred during the defects liability period, provided that they have not been caused by the occupier. It is vital that the buildings should be carefully inspected at the end of the liability period, because if there are any obvious defects at that time that the architect does not identify, it may well be assumed that the architect was prepared to accept the building subject to those defects.

The importance of inspecting the site and its immediate environs on the handover date should not be overlooked. If during the building contract any damage has been caused to adjoining property, damage to boundary walls and fences is not unusual, then the contractor must remedy it. Inspection of the roads, footpaths, kerbs, grass verges etc. immediately adjoining the site is carried out to see that the contractor remedies any damage, otherwise the highway authority might subsequently ask the developer to bear the cost of any remedial works.

The architect should produce as 'built drawings' a building manual and maintenance schedule to assist the occupier by giving a comprehensive schedule and description of all components (taps, locks, fastenings, sanitary ware etc.) that might need replacing at some future date, together with recommendations for regular maintenance work to preserve the building fabric. Similarly, manuals and operating instructions for services are provided by the services engineers.

Where an occupier is not taking possession immediately, the developer is responsible for the vacant building and a programme of regular cleaning and maintenance should be instigated. Whatever physical arrangements are reasonable and necessary to protect the property against vandalism should be made. An example is in a shopping development, where un-let shop units will have a neat hoarding put across the frontage immediately before the handover date. Consideration is often given to the issue of employing security guards or patrols. Adequate insurance cover should be in place to give protection against fire and loss due to the damage of property. Public liability insurance should also be arranged to protect against claims from injured third parties.

Monitoring construction progress and costs

The project manager's objective is to produce the building on time and within budget for the developer client. Therefore, it is important to examine how the project manager reports to the developer on construction progress and cost. Any delays in completion or increase in costs will affect the profitability of the development, therefore, it is essential that a developer is kept regularly informed on progress and cost. The developer will need regularly to update the cash flow appraisal prepared at the initial evaluation stage (see Chapter 3) to assess profit.

Every project manager will have their own method of reporting but it is important to agree with the developer at the outset the information required. The best method of reporting uses charts and graphs to compare actual progress and cost against the original estimates. The starting point should be the appraisal used at the site acquisition stage. Actual costs and progress are compared against the estimates made at the time of acquisition. This means the developer can easily identify changes in costs and progress, instead of reading through pages of written information. The charts usually have written comments on them giving reasons why costs have increased/ decreased or why site progress is behind schedule. Once the developer has assimilated the information in the charts and graphs further questions can be asked of the project manager as to the reasons behind identified increases in cost or delays in process. In particular, it should be clearly understood who is able to authorise the expenditure of money. Every member of the project team must know whether they are able to spend money and, if so, what authorities they must obtain.

Typical reporting methods include the following:

1. **Bar chart** The bar chart is a calendar showing the development programme in weeks or months. The programme is divided into tasks and the period during which each of these is to be carried out is shown on the chart. An example of such a chart is shown in Figure 6.1. The chart includes pre-contract activities as well as the contract programme, which are equally important to monitor as any delay will impact on the start of construction. This shows how crucial it is for the project manager to be involved in the development process from the beginning. The chart indicates when each task is to start and finish. It shows how the tasks overlap and the work that should be in hand at any time. From time to time the programme, and the bar chart, may need to be amended but a comparison of what has been achieved against what the chart shows gives the developer and project manager a simple yet instant test of progress.

   The bar chart can be used to indicate when information or decisions are needed by the project manager from the developer and by the contractor. This is vital as lack of information or instructions is one of the main causes of delay. The bar chart demonstrates that delay in one activity can affect the whole programme. Once a delay is identified it is important for the project manager to advise the developer what affect it will have on the overall programme and how time can be made up in other activities. It is vital for the project manager to issue the bar chart

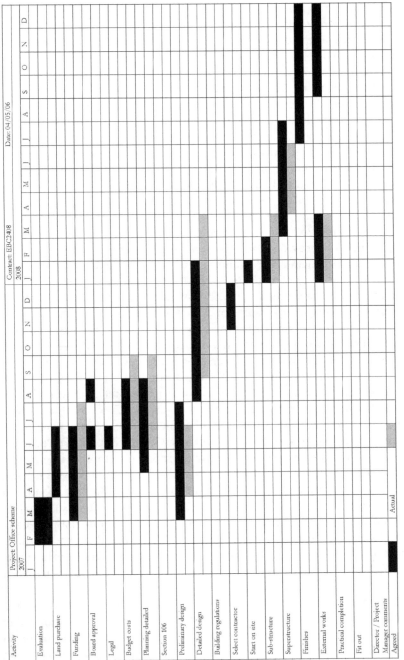

Figure 6.1   Overall development programme – bar chart

to the entire professional team, so that each member of the team can identify the target dates they have to work to.

Once the contractor is on site, the bar chart may be substituted with the contractor's own bar chart, which identifies the timescale for each trade involved on site. It is important that the project manager receives the contractor's bar chart regularly so that the overall bar chart can be updated and amended as necessary. The developer may not need to know the progress of each individual trade on site and it is often sufficient to break down the contractor's programme, into substructure, superstructure, finishes and external works. However, the project manager must be able to produce the contractor's bar chart at any time, as in some cases the developer will need to know the detailed programme. For instance, the developer may need to know when the area in the building identified as a show suite is ready (see Chapter 8).

Another method of monitoring progress and highlighting the importance of providing information and decisions is to prepare a chronological timetable of events. However, the bar chart is the most instant way of comparing progress against original estimates.

2. **Cash flow table and graph**   A cash flow table and graph is prepared by the project manager, examples of which are shown in Figures 6.2 and 6.3. The purpose of the cash flow table in Figure 6.2 is to estimate the developer's flow of cash payments throughout the development period. The developer can use this to prepare a cash flow appraisal, which can be regularly updated throughout the development. The importance of the cash flow has already been discussed in Chapter 3. The table can be represented as a graph as shown in Figure 6.3. The combination of the table and graph can provide another means of checking progress by comparing actual with estimated payments. However, it is less effective at measuring progress than either the bar chart or a development timetable. Estimates of cash flow often have to be revised and there is a danger that they do not really highlight problems of delay until the last months of the contract.

3. **Financial report**   The project manager's financial report may typically look like the example shown in Figure 6.4. It is based on the quantity surveyor's cost reports and payments already made to the contractor as certified by the architect. It enables the developer to identify variations in costs throughout the contract.

The project manager should advise the developer of the reason for the cost variation. Any variation in cost from the original contract value may be due to a claim from the contractor, architect's instructions to the contractor or variations required by the developer. We have already

Project: Office scheme 2007 — Contract: EBC2408 2008 — Date: 04/05/06

Fees (000's)	budget	Total to date	J	F	M	A	M	J	J	A	S	O	N	D	J	F	M	A	M	J	J	A	S	O	N	D
						2007												2008								
Architect	450	336							30	30			30		20					12						
Structural Eng'r	180	90							24	22			22		12					12						
QS	180	90							22	24			22		12					12						
M&E Eng'r	176	68							34																	
Proj Man	180	80								40			44							20						
Acoustic																									
Landscape																									
Party Wall																									
Right of light																									
Site surveys	6	6																							
Ground surveys	16	30																							
Planning	6	6								18																
Building Regulations	44	26																							
Others																									
Demolition																									
Enabling																									
Main contract	9000	1670						1396	1598	922	948	1056	636	450	141	216				276						
Statutory authorities																									
Fitting out																									
Budget total	10198	2698						918	810	852	1080	1170	1194	720	430	216			56	270						
Actual total	10398	2262						1506	1598	1056	948	1056	754	450	220	216			56	276						
Building/Project manager comments																									

Figure 6.2 Cash flow table: fees/construction

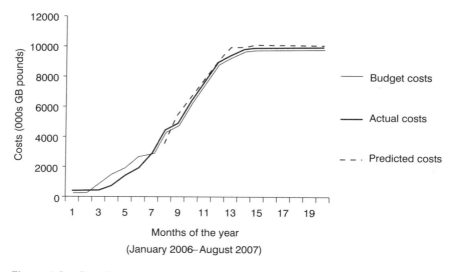

Figure 6.3   Cash flow graph: fees/construction

examined the circumstances under which a contractor can make a claim for additional costs. Claims may be based on the inadequacy of the drawings and/or the Bill of Quantities. In addition, they may be based on delays in the architect issuing drawings or instructions. The project manager must ensure that claims and variations are kept to a minimum if costs are to be kept in budget and should monitor the activities of the architect and ensure they keep to their drawing schedule. The project manager together with the quantity surveyor must advise the developer of the cost of any variation proposed to ensure the developer is aware of the implications. No revision should be made without justification. The developer must know why cost estimates have to be revised. The project manager must maintain a scrutiny of costs and question any decision which has a cost implication.

4. **Checklist**   Most project managers prepare a checklist of the main activities throughout the development. An example of such a checklist is shown in Figure 6.5. The checklist defines the main activities applicable to the particular development, some of which will require approval by the developer. It should highlight information required by the project manager from the developer. It is essential for the project manager to identify the decisions needed by the developer and by what date. Developers should know if the progress of a development is being held up because a decision is required of them, and the implications of any

	Project Office Scheme	Contract EBC 2408						Date 04/05/06
	Board approval date	Revised report date		Revised report date		Revised report date		Director/Project manager comments
	15/9/05	10/05/06	+(-)	10/05/06	+(-)		+(-)	
	8/1/05	8/1/05		8/1/05				
Site start								
p/completion	21/12/05	28/1/06	– one month	28/1/06	– one month			Agreed extension of time see memo
Net lettable area	38,250	38,250		38,250				Client to measure in July
Building contract value	9,000,000							For cost breakdown see QS report no. 6.
Demolitions								
Enabling works								
Sub structure		+106,068		+106,068				Increased cost of piling due to ground conditions.
M&E Services		–13,480						Saving due to design change.
Finishes		7,906,068		7,882,388				
Externals works								
Preliminaries								
Inflation								
Contingencies	200,000	200,000		200,000				
Statutory services								
Tenant works								30,912 of contingency not expended
Claims (unsettled)								
Instr. (not priced)								
Pending instructions								
Others agreed ext. of time		46,500	+46,500	46,500	+46,500			
Clients variation				30,000	+30,000			Delay caused by late receipt of drawings from structural engineers. Clients variation to finishes needs to be finalised.
Total	9,000,000	9,156,508	156,568	9,169,088	–169,088			

Figure 6.4 Financial report: building costs

Projects Office Scheme		Contract no. EBC 2408			Date 04/05/06		
Month		May 2006		June 2006			
	Approved by	Project manager	Client	Project manager	Client		Director/project manager comments
Clients brief		✓	✓	✓	✓		
Select/appoint architect		✓	✓	✓	✓		
Select/appoint QS		✓	✓	✓	✓		
Select/appoint structural engineer		✓	✓	✓	✓		
Select/appoint M&E		X	X	X	X		All agreed except one item
PM appointment		✓	✓	✓	✓		All agreed except arch.
Deeds of collateral warranty		X	X	X	X		
Letters of intent							
Development/feasibility appraisal			✓		✓		
Site boundary/ownership agreed			✓		✓		
Appoint rights of light							
Appoint party wall							
English heritage agreement							
Summary of funding documentation			✓		✓		
Funds/banks surveyors approval							
Tenant requirements							
Agreed contract programme		✓	✓	✓	✓		
Planning dgs/application		✓	✓	✓	✓		
Section 106 agreement							
Freeze design/stage report		✓	✓	✓	✓		
Building regulations application		✓	✓	✓	✓		
Certificate of readiness		✓	✓	✓	✓		
Summary of insurance requirements		✓		✓	✓		
Summary of building contract conditions		✓	✓	✓			
Appoint building contractor		✓	✓	✓	✓		
Agreement with statutory undertakers		✓	✓	✓	✓		
Signed receipt for maintenance manuals		X	X	X	X		
Client decisions (major items)		X	X	X	X		
Finishes board - urgent		X	X	X	X		Meeting arranged 16/06/06 to discuss

Figure 6.5   Main activities checklist

delay in the decision. The developer/project manager relationship is a two-way one and both should ensure the other is kept fully informed at all times.

The above mentioned methods of reporting are typical but every project manager will have their own method. Whilst regular reporting on progress and cost is a way of keeping the developer informed, it also provides the project manager with an essential tool. If the developer insists on regular reporting in the manner shown above, then the project manager will know whether the aims are being achieved. It will bring into sharp focus the targets that need to be achieved and the problems that need to be tackled.

The project management of a development through the construction process, whether carried out by the developer or through the appointment of a project manager (or other professional) is about teamwork and motivating the team to work together. Problems must be sorted out before team members resort to a blame culture and become entrenched. The project manager must anticipate delays by ensuring constant communications with the professional team and the contractors. Contracts and paper communications should not be relied on: there is no substitute for personal contact. Overall, project managers should fulfil their role efficiently, constantly considering cost and time. They need the ability to lead and motivate the professional team and the contractor. This, again, shows that the property development process is all about the interaction between people.

Discussion points

- What are the main issues that developers need to consider during the construction phase of a project?
- What are the major risks the developers faces during this stage?

6.4 Public–private partnerships

Collaboration between public bodies, such as local authorities or central government, and private companies is known as a public–private partnership (PPP). Public developments are now able to consider PPP as a method of procuring buildings and infrastructure. In the public sector, there are three main procurement approaches: PPP, design and build and prime contracting. The rationale for PPP is that private companies are more efficient and better managed than public bodies. In bringing the public and private sector together, the aim is that the business community's management and financial skills will lead to better value for money for taxpayers. The Private Finance Initiative (PFI) was created the early 1990s. Governments and local authorities traditionally paid private contractors to build roads, schools, prisons and hospitals out of tax money. Under PFI, contractors pay for the construction costs and then rent the finished project back to the public sector. This enables government to get new hospitals, schools and prisons without raising taxes. The contractor is allowed to keep any cash left over from the design and construction process, in addition to the 'rent' money. Critics say that governments are mortgaging the future and that the long-term cost of paying the private sector to run these schemes is more than it would cost the public sector to build them itself.

PFI is in its infancy for hospitals and schools but it is a well-established way of paying for new roads and prisons. In 2007, there were eight new, private prisons – with more on line – and major road schemes like the Thames crossing and the Birmingham relief road financed through PFI. The complex nature of PFI contracts and the political obstacles involved in getting controversial schemes like the London Underground PPP launched mean that progress in some areas has been slow. It has been estimated that trade in public services could ultimately net the private sector an extra £30 billion a year. This breaks down into approximately £20 billion in central government contracts, £5 billion in education and £5 billion in local authority contracts.

PFI has broadened the concept of public–private co-operation. If privatisation is a takeover of a publicly-owned entity, PPP is more like a merger, with both sides sharing the risks and seeing the benefits. With health and education accounting for 13 per cent of Britain's GDP, the rewards for industry of opening up the public sector to private finance are huge. However, the government is not clear how far it wants to go in these areas in the face of opposition, and critics argue that taxpayers will end up paying for PPP. With Fazackerly prison in Liverpool, the initial cost of the project was claimed to have been paid back within two years, leaving 23 years of pure profit from the construction. There are claims that some PFI projects have been substandard, with private companies taking shortcuts to maximise profits. Another PFI criticism is that firms make their profits by reducing employees' wages and benefits. However, PPP supporters maintain that some hospitals and schools would not be built if it was not for private finance and assert that PFI will lead to increased quality in public services. Performance-related penalties, now part of most PFI contracts, will ensure improvement in standards. PFI is fast, effective, and in the short term, an economic way of getting new projects built. The biggest UK hospital-building programme is currently underway because of PFI and the government is gearing up for a large increase in private involvement in public services in the future, with local authorities being directed towards PPP.

6.5 Partnering

During the 1990s, the Latham and Egan reports (Latham, 1994; Egan, 1998) identified the inefficiency inherent in an adversarial UK construction industry. In addition, a UK government report (National Audit Office, 2001) showed 73 per cent of their projects were over budget and 70 per cent ran over time. In particular, Egan's report set a challenge for change and improvement. Clients, designers and contractors have taken up the gauntlet and developed different types of business relationships, among which is partnering. Under this type of arrangement, parties to a contract work towards agreed goals that will benefit all concerned. Partnering thrives in an atmosphere of trust and openness and fails when co-operation is absent. In summary, partnering is a business relationship for the benefit of all parties – built on trust, openness and respect. Whilst the contract establishes the legal relationship between parties, partnering establishes the working relationship. Some refer to it as the traditional way of doing business where a person's word was their bond. Partnering can be used on small, large or complex projects and is promoted as a win–win way of doing business for all parties. Partnering is based on an ethos of teamwork and in the case of construction projects goals are achieved through a teamwork approach to:

- design control and efficiency;
- minimising pre-construction budgeting and approvals periods;
- maximising efficiency of the construction period (and completion dates);
- problem-solving co-operation;
- cost control reporting and reconciliation;
- agreed conflict or dispute resolution procedures.

For partnering to succeed, a number of key principles have to be adopted by the parties – for example, a commitment to, and value placed on, a long-term business relationship, thereby ensuring a willingness to work towards longer-term goals, such as a reduction in project times and improvement in building performance (quality). In the partnering organisations there is a requirement to develop an environment for long-term profitability and to encourage innovation. In addition, partners commit to improved project buildability and a lowering of project costs through the process of value management. The establishment of project organisational structures and clear lines of communication are required to reduce conflict and disputes and ensure successful outcomes. Advocates claim that the successful outcome is a project constructed in less time, costing less and of higher quality than would otherwise be realised through traditional procurement routes. With a partnering approach, it is argued, developers and contractors are able to define the project better and identify risks prior to commencement to avoid time delays and cost overruns and the subsequent poor relations between parties. In summary, the key elements of partnering in construction are:

- commitment from top management – a jointly developed partnership charter is not a contract but a commitment;
- equity – all stakeholders' interests are considered in establishing mutual goals and there is a commitment to win–win thinking;
- trust – teamwork without trust is not possible, and personal relationships are developed to build trust and understanding about each stakeholder's risks and goals;
- development of mutual goals/objectives – at partnering workshops, mutual goals and objectives are established and the means by which to meet them are identified;
- continuous evaluation – to ensure implementation, stakeholders agree to a plan for periodic joint evaluation based on mutually agreed goals;
- timely responsiveness – saves money and can stop a problem growing into a dispute. Methods of discussing issues are discussed prior to project commencement to reduce conflict risks.

Partnering workshops are set up at the commencement of the project and involve all team members. The workshop, a planning session to establish the partnering tools and to problem solve critical design/construction issues, typically takes two days to complete. The project charter is agreed to be a most visible tool, and sets out the team performance goals and mission for the project. It is periodically reviewed to ensure that it has continued appropriateness. It is vital to agree criteria to measure whether the partnering is successful. The next stage is, therefore, to agree a team report card and the benchmarks for measuring success. The third and a most important tool developed at the initial workshop is the issue resolution process. The process should encourage communication and creative problem solving. Roles and responsibilities are defined and communication points are agreed. Without a commitment to principle-centred leadership from management, partnering will flounder. Follow-up meetings are held every 3 or 6 months during the project and it is in the follow-up meetings that partnering is actually delivered and results produced. Similarly, at the end of the project, an evaluation should be undertaken so that lessons can be learned by all stakeholders for future projects.

Reflective summary

The developer's aim during the construction process is to produce a good quality building on time and within budget. The choice of procurement is critical and is dependent on project size and complexity as well as the developer's attitudes to risks. There are three main types of building contract available: traditional JCT contract, design and build, and management contracting, although there are many variations on each depending on the exact contractual arrangements and the role of the professional team. Each contract has its main advantages: the traditional contract for its flexibility; the design and build for cost control; and management contracting for speed. Disillusionment by developers with the traditional contract has led to an increased use of the design and build contract where the contractor is responsible for design and construction, avoiding the problem of cost increases due to the late production of drawings by the architect. To overcome the problem of quality control with the design and build contract, the appointed contractor has taking over the appointment of the developer's chosen architect through a novated contract arrangement. Whichever method is used, the success of the building contract, in terms of achieving the aims above, relies on good control, leadership and firm early decision making by the developer and the project manager. The professional team and the contractor should be motivated towards the same goal, resolving any conflicts and problems before they arise. One factor impacting considerably on the construction phase in the UK is the shortage of labour, particularly in plumbing, electrical and plastering trades. In the public sector PPPs and PFI are affecting procurement. The advent of partnering is also affecting some developments with distinct advantages for long-term working relationships.

Case study

This case study is a good example of the use of a design and build contract, in a varied form, on an out-of-town retail park. The developer had pre-let and forward-funded the scheme, and by the time a decision had to be made on the form of building contract the remaining risk to the developer was the construction cost and time. Having experienced excessive claims from contractors previously, the developer was disenchanted with the traditional JCT contract but had no experience of the design and build procurement method. This case study examines how the standard design and build contract was varied and administered, focusing on the construction phase of the development.

Background

The developer specialises in retail development, both town centre and out-of-town schemes, although they have undertaken limited office development. The development was a greenfield site on the edge of a town on an 'A' road, adjacent to an existing major food superstore. The developer obtained detailed planning consent for a scheme comprising 7,150 m² of non-food retail and a 1,330 m² food supermarket, and secured pre-lets on the entire scheme, which is usual with such out-of-town retail schemes. The tenants include a supermarket, a garden centre and some of the smaller DIY and home furnishing retail multiples. With the pre-lets the developer was able to secure forward-funding with a fund.

The contract and role of the professional team

The professional team comprised the architects, the quantity surveyors and a structural engineer and was appointed on the basis of adopting the traditional JCT route. However, at the point where a decision had to be reached on the type of building contract, it was collectively decided that the design and build route would be adopted instead, due to recent bad experiences previously by all parties.

To retain some control over the quality of the design and construction the developer wanted the architect, the quantity surveyor and the engineer to be involved in the scheme throughout the construction period. Therefore, a variation on the standard design and build contract was selected. The developer would novate the contracts (letters of appointment) of both the architect and engineer to the contractor, that is to say, the contractor would become the client of both professionals on identical terms to those contained in their letters of appointment with the developer. The contractor was then responsible for undertaking the detailed design and construction of the scheme with the appointed professionals. Furthermore, the developer appointed the same architect to be directly responsible for quality control. At this stage the architect was responsible to two clients and a potential conflict of interest existed, so it was decided that one partner in the practice would work for the developer and another partner would work for the contractor.

Once the design and build procurement method had been decided the architect was instructed to prepare a performance specification incorporating various tenants' specifications in great detail, including a specification of all materials to be used. The tenants would be responsible for their own fit-out but the developer's contractor would incorporate items such as fire protection and toilet facilities as part of the main contract. It is usual where the developer bears the cost of such tenant's facilities that they become landlord's fixtures and fittings under the terms of the lease, and are taken account of for rent review purposes.

Initially, the performance specification was sent to a contractor and the price submitted to the developer was considered too high. A second contractor was invited to tender; and, although they came up with a lower price, the fund would not consider them due to concerns about their financial standing. The first contractor was appointed on a negotiated 'JCT 1998 Design and Build Contract with Contractor's Design', but on the basis of novated contracts with both the developer's architect and engineer. A programme of 6 months was written into the contract, although it was accepted by all parties that this was tight. The supermarket unit became a separate contract between the retailer and the contractor as the retailer had acquired a long leasehold interest in their site from the fund.

At the time of negotiation, tender market conditions favoured the developer's position as competition was fierce for the few building contracts available. This enabled the developer to place most of the risk of increases in costs and delays onto the contractor. Note that a developer will always aim to minimise risk at every opportunity during the development process whilst maintaining a reasonable profit level. Under the terms agreed on this contract the contractor would only be able to claim an extension of time if any delay was caused by the developer's variations to the contract. The standard clauses such as allowing the contractor to claim for delays caused by bad weather or changes in legislation were omitted from the contract. Furthermore, it was agreed that any variations made by the tenants would be accommodated by the contractor only if they did not cause any delay on the main contract.

The pre-contract period took 6 months, with much time taken up by the issue of collateral warranties that had to be agreed on behalf of the fund with the professional team and the contractor (together with subcontractors). Collateral warranties involved the developer in lengthy negotiations with the solicitors acting for the fund and the professionals and contractors.

The quantity surveyor, who had a long-standing working relationship with the developer, was appointed as the 'employer's agent' (developer's representative) for the contract. Usually, this would involve the quantity surveyor as both the contract administrator (in place of the architect who traditionally performs this role under a traditional JCT contract) and the project manager. In fulfilling such a role the quantity surveyor would then be responsible for appointing the professional team and agreeing their deeds of collateral warranty; chairing and minuting all project meetings; and preparing all interim valuations and the final account. However, on this particular scheme what, in effect, happened was that the architect chaired and minuted the meetings and the quantity surveyor undertook a normal role. The funds were represented at project meetings by their structural engineer, quantity surveyor and building surveyor.

Conclusion

The contract took a little over 7 months and was 6 weeks overdue. As the original programme was tight the developer agreed to a compromise in respect of the liquidated damages to be paid by the contractor. Under the terms of the contract the contractor was due to pay £55,000 a week, however, this was reduced to £2,200 for the first four weeks of the claim and £55,000 for the last fortnight. Overall, the developer was pleased with the financial outcome of the contract as the scheme was more or less on budget. The developer was happy with the quality of the scheme and plans to use design and build in future. However, on reflection, keeping the two roles of the architect separate was not easy and is not the ideal way of arranging such appointments. Currently, this variation of the design and build contract, with novated professional appointments and with quantity surveyors acting as the developer's representative, is gaining favour. Some developers use design and build on all their developments.

CHAPTER

Market research

7.1 Introduction

The importance of sound market research is often underestimated in the property development process, but in reality conducting thorough market research has the potential to make or break a successful property development. Professional property developers acknowledge that well planned and executed market research will substantially increase the likelihood of success although there will always remain some information that is either unknown or unknowable. The fast-growing area of property market research continues to embraces different specialisms including information and database services, strategic and site-specific analysis, forecasting and portfolio analysis.

Site-specific and strategic analysis are used by developers and investors to help assess the viability of individual projects and/or to inform the company's long-term property strategy. Analysts can provide invaluable insight about the fundamental demand and supply factors underlying market conditions. In addition, market research can pinpoint risks and opportunities that might not be readily evident from current transactions in the market. This chapter discusses the importance of market research and how it can assist the property developer. Also discussed are the differences between land types that further complicate direct market comparisons. Prior to commencing the market research the nominated land type should be identified early in the analysis, which in turn will assist the overall market research process.

(Some of the information in this chapter is based on Chapter 9 ('Market Analysis') in the *Valuation of Real Estate* published by the Australian Property Institute, 2007, which is based on *The Appraisal of Real Estate* published by the Appraisal Institute, 2001.)

7.2 Background to conducting market research

Generally speaking, market research means different things in different circumstances. For example, the terms 'market research' or 'market analysis' are used broadly in economics but have a more specific meaning when related to property development. Overall, market research or analysis is the identification and study of the market for a particular economic good or service and it can be considered at two levels:

1. from a broad market viewpoint, without a specific property as the focus of the study;
2. from the perspective of the real estate market in which a given property competes.

Although there is a logical continuum from the general to the specific, market research applied to a specific property should not be confused with general market research or related studies. This is an important distinction.

When undertaking market research and gathering data, it is important to interact with many market stakeholders and participants. Key sources are those involved or likely to be involved in property transactions that can be relied upon for comparison purposes. In addition, other sources may provide only general or broad information about the local economy as supporting information. It is important to understand how the interaction of supply and demand affects the property's value. Through the investigation of recent property transactions, properties for sale in the competing market (although not necessarily in the same geographical market) and the behaviour of market participants, it is possible to examine supply and demand relationships and investigate the reasoning behind the prices paid and the prices accepted. This will assist in the interpretation of market attitudes toward current trends and anticipated changes. If current market conditions do not indicate adequate demand for a proposed development, the market research may identify the point in time when adequate demand for a project will likely emerge. Therefore, market research helps to forecast the timing of a proposed property development and the amount of demand anticipated over a particular period of time.

With regards to property development, market research provides a sound basis for determining the highest and best use of a property. In other words, an existing or proposed improvement under a specified use may be put to the test of maximum productivity but only after it has been demonstrated that an appropriate level of market support exists for that use. In-depth market research goes much further in specifying the character of that support. For example, such studies may determine key marketing strategies for an existing or proposed property, address the design characteristics of a proposed development or provide estimates of the actual proportion of the overall market that the property is likely to capture, and its probable absorption rate.

To measure the market support for a proposed property development, the researcher must identify the relationship between demand and competitive supply in the subject property market, both currently and in the future. This relationship indicates the degree of equilibrium or disequilibrium that characterises the present market and the conditions likely to characterise the market over the forecast period. The market value of a proposed property development is largely determined by its competitive position in its market. Familiarity with the characteristics and attributes of the subject property will enhance the ability to identify competitive properties (supply) and to understand the comparative advantages and disadvantages that the subject offers potential buyers or tenants (demand). With an understanding of economic conditions, the effect on property markets and the momentum of these markets, it is then possible to identify and better understand the externalities affecting the proposed property development.

7.3 Basic market research concepts

Identification of the market and segmentation

At the beginning of the market research process it is clearly important to identify the proposed property development and the property market in which it competes. These two tasks may be considered complementary. Analysing the characteristics and attributes of the subject property helps to identify competitive properties that constitute the applicable market. Defining the property market for the subject property clearly enhances the understanding of how externalities affect the proposed development. By conducting market analysis, the researcher breaks down a specific property market into consumer submarkets or market segments, and then disaggregates the proposed property development from other types of properties.

The actions of property market participants are prompted by their expectations about the uses of a property and the benefits that property will offer its users. Market segmentation differentiates the most probable users of a property from the general population by their consumer characteristics. The activity of individual market participants in a property market focuses on a real estate product and the service it provides. Product disaggregation, therefore, differentiates the subject property and competitive properties from other types of properties on the basis of their attributes or characteristics.

A market segment is delineated by identifying the market participants likely to be involved in transactions related to the proposed property development and the type of real estate product and service it provides. Product disaggregation includes both the subject property and competitive and complementary properties. Therefore, market research combines market segmentation and product disaggregation. The characteristics of a proposed property development and its market area that are investigated by a researcher in the process of delineating the market are listed in Table 7.1.

Discussion point

Market research can be defined in various ways. What are the main variables that define markets?

Demand

Overall demand reflects the needs, material desires, purchasing power and preferences of consumers. Demand analysis focuses on identifying the potential users of a proposed property development, i.e. the buyers, tenants or customers it will attract. For each particular type of property, demand analysis focuses on the end product or service that the property provides. For example, a demand analysis for a proposed office development would attempt to identify businesses in the area that occupy office space and their space or staffing needs. A demand analysis for retail space would attempt to determine the demand for retail services generated by potential customers in the market area.

Demand analysis for residential and retail markets specifically investigates the households in the subject's market area (note that a household is defined as a number of related or unrelated people who live in one housing unit, where even a single individual may constitute a household). In addition

Table 7.1   Market segmentation process

To identify a specific real estate market the following factors must be investigated:

1. Proposed property development (e.g. retail shopping centre, office building).

2. Property features such as occupancy, customer base, quality of construction, and design and amenities.

    a. occupancy: single-tenant or multi-tenant (residential, residential unit, office, retail).
    b. customer base: the most probable users. Data on population, employment, income, and activity patterns is analysed. For office markets, the customer base reflects the space needs of prospective companies leasing office units. For commercial markets, data is segmented according to the likely users of the space. For retail markets, the clientele that the prospective tenants will draw represents the customer base. For residential markets, data is broken down according to the profile of the likely property owner or tenant.
    c. quality of construction (class of building).
    d. design and amenity features.

3. Market area: defined geographically or by location. A market area may be local, regional, national, or international in scope. It may be urban or suburban; it may correspond to a district or neighbourhood of a city. Retail and residential market areas are often delineated by specific time-distance relationships. Refer to Figure 7.1 for a simple overview of a market defined by geographic location where the black dots (e.g. housing) are located in the catchment area for the subject property (e.g. shopping centre).

4. Available substitute properties i.e. equally desirable properties competing with the property development in its market area, which may be local, regional, national, or international.

5. Complementary properties i.e. other properties or property types that are complementary to the property development. The users of the subject property need to have access to complementary properties, which are also referred to as support facilities.

to the number of households in the market area, this research focuses on the disposable income or effective purchasing power of these households and the ages, gender, preferences and behavioural patterns of household members. Often this type of research refers to the government population census or a 100 per cent sample, which divides the population into different census tracts and collects detailed demographic information. An example of identifying census tracts based on proximity to the proposed property development is shown in Figure 7.2, where the location of transport routes and natural barriers (e.g. rivers) must be considered.

The demand for housing and most retail space is projected on the basis of growth rates in population, income and employment levels. The key

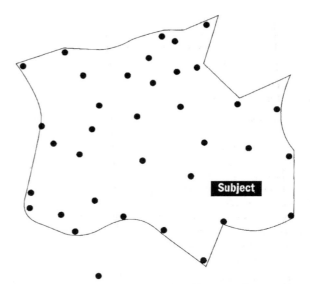

Figure 7.1 Defining a market area via geographical location (Source: Fanning, 2005)

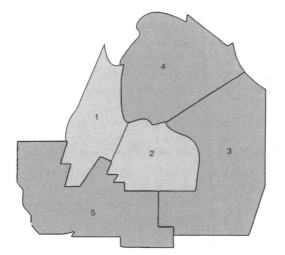

Market area	Census tracts
Primary	1, 2
Secondary	3, 4, 5

Figure 7.2 Defining a market area via household census tracts (Source: Fanning, 2005)

points discussed below can be especially useful in understanding demand projections for a particular land use:

- The rate of household formation varies significantly with income and age (cohort) groups in the existing population; this rate is even more sensitive to migration. Estimating the number of households in an area by dividing the total population by the average household size may result in considerable error.

- Although average or median income is generally projected in current income levels by using the local currency, real income is not stable and may either decrease or increase. Therefore, income projections based on the local currency will probably reflect future, inflated dollars.
- Population projections for small areas may be published by estate agencies and market research firms but such projections can be misleading. Therefore, the researcher should also consult projections for the overall metropolitan area. The availability of property for development and the adequacy of the infrastructure in the subject area will help determine how much of the overall growth projected will go into a particular area.

Important factors in demand analysis for a proposed office development

- Employers who use office space; current and estimated future staffing needs.
- Average square metre areas of office space required by an office worker. Note that requirements vary according to the category of work, the rank of the office worker and the location of the office in the suburbs or the central business district.
- Vacancy rate for the specific class of office building.
- Pressure to upgrade placed on higher grade buildings or decreased demand for lower grade buildings.
- Land-use patterns and directions of city growth and development.
- Accessibility (transportation facilities and highway systems) and cost of transportation.
- Factors that affect the appeal of the office building (quality of construction, management and tenancy) and the availability of support facilities (shops, restaurants and recreational centres).

Important factors in demand analysis for a proposed retail development

- Population of catchment area(s) – size and numbers of households, rate of increase or decrease in household formation, composition and age distribution of households.
- Per capita and household income (mean and median).

- Percentage of household income spent on retail purchases and percentage of disposable income (effective purchasing power) spent on various retail categories.
- Rate of sales retention in the catchment area.
- Required volume of sales for a retail facility to operate profitably and existing sales volume per square metre.
- Retail vacancy rate in the market.
- Percentage of retail purchases captured from outside the catchment area.
- Land-use patterns and directions of city growth and development.
- Accessibility (transportation facilities and highway systems) and cost of transportation.
- Factors that affect the appeal of the retail centre (image, quality of goods and tenant reputation).

Important factors in demand analysis for a proposed industrial development

- Proximity to raw materials.
- Exchange capability (e.g. currency values and trade barriers).
- Area employers who use industrial space; current and estimated availability of skilled and unskilled labour.
- Land-use patterns and directions of city growth and development.
- Accessibility (transportation facilities and highway systems) and cost of transportation.
- Employment in manufacturing, wholesale, retail, transportation, communications or public utilities.
- National and regional economic growth that affects local demand.
- Overall employment growth.
- Retail sales (applicable in market analysis for retail storage and wholesale distribution properties).
- Goods flows by transport type (e.g. truck, rail, water, air) and product type (e.g. high or low bulk).

Note that demand in industrial property markets is generally more limited in comparison to demand in residential or commercial markets.

Important factors in demand analysis for a proposed residential development

- Population of the market area – size and number of households, rate of increase or decrease in household formation, composition and age distribution.
- Income (household and per capita).
- Employment types and unemployment rate.
- Percentage of owners and renters.
- Financial considerations such as savings levels and lending requirements (e.g. interest rates on mortgages, loan-to-value ratios).
- Land-use patterns and directions of city growth and development.
- Factors affecting the physical appeal of the neighbourhood, e.g. geography and geology (climate, topography, drainage, bedrock and natural or man-made barriers).
- Local tax structure and administration, assessed values, taxes and special assessments.
- Availability of support facilities and community services (cultural institutions, educational facilities, health and medical facilities, fire and police protection).

Discussion point

Different land uses have different supply and demand drivers.
Consider the differences between these drivers when comparing different land uses.

Competing supply

Supply refers to the production and availability of the real estate product. To analyse supply, the researcher must compile an inventory of properties that compete with the proposed property development. Competitive properties include the stock of existing units, units under construction that will enter the market and projects currently in the planning pipeline, for which important factors are listed in Table 7.2. Care must be exercised in developing and analysing data on proposed or announced projects because some may not ultimately be constructed. The researcher must also determine the number of units lost to demolition and the number added or removed through conversion, although the degree of utilisation of these properties should also be considered. Data may be gathered in various ways:

Table 7.2   Important factors in the analysis of the supply of competing properties

---

- Quantity and quality of available competition (standing stock).
- Volume of new construction (competitive and complementary) – property developments in planning and under construction.
- Availability and price of vacant land.
- Costs of construction and development.
- Properties currently on the market 'for sale' (existing and newly built).
- Owner occupancy versus tenant occupancy.
- Causes and number of vacancies.
- Conversions to alternative uses.
- Special economic conditions and circumstances.
- Availability of construction loans and financing.
- Impact of building codes and bylaws, zoning, town planning controls and other regulations on construction volume and cost.

- field inspection, e.g. vacant sites;
- review of building permits (issued and acted upon), cadastral maps, town plans and surveys of competitive sites;
- interviews with developers and government planners;
- review of newspaper articles and media coverage.

Market equilibrium

Over the short term, the supply of property is relatively fixed and prices are responsive to demand. If demand is unusually high, prices and rents will start to rise before the construction of new property developments can commence. The completion of a property development may lag considerably behind the shift in demand. In other words, disequilibrium generally characterises property markets over the short term.

Theoretically, the supply of and demand for real estate move towards a point of equilibrium over the long term. However, in reality this point is seldom achieved. In some markets, such as those characterised by a narrow specialised economy such as computer technology, supply often responds slowly to changing demand conditions. Even when there is an excess of supply, account must be taken of the fact that property developments currently under construction will generally be completed. Therefore, additional stock will continue to be added to the existing surplus and cause greater disequilibrium. A decline in demand may also occur while new property developments are being completed, further exacerbating the level of oversupply.

7.4 Trends in market activity

Analysts and market participants describe the activity of real estate markets in a variety of ways. An active market is a market characterised by growing demand, a corresponding lag in supply and an increase in prices. An active market is also referred to as a seller's market because the sellers of available properties can insist upon higher prices. A depressed market is a market in which a drop in demand is accompanied by a relative oversupply and a decline in prices. A depressed market is also referred to as a buyer's market because purchasers have the advantage with a stronger negotiating position.

Other terms applied to markets are subject to interpretation. For example, markets are sometimes characterised as 'strong' or 'weak'. In such cases, strong markets may reflect both high demand and increasing price levels or a large volume of transactions. Weak, or 'soft', markets may be identified by low demand and declining price levels. Other loosely defined terms include 'broad' and 'narrow' markets, 'loose' and 'tight' markets, and 'balanced' and 'unbalanced' markets. Note, however, that care must be taken in describing markets in this way, since supply and demand does not always act as expected. For example, supply may fail to respond to increasing demand because the rate of demolition may exceed the rate of newly completed development. In this example, prices will continue to rise. On the other hand, supply may outpace rising demand because of a glut of existing properties on the market, and prices will decline.

Above all, the activity of the real estate market is cyclical. In a similar manner to the business cycle, the real estate cycle is characterised by successive periods of expansion, decline, recession and recovery. However, the real estate cycle is not synchronised with the business cycle. Real estate activity responds to both long-term and short-term stimuli. Simply explained, the long-term cycle is a function of changes in the characteristics of existing employment, population, income and shifts in consumer preferences.

Levels of market analysis

The principles of market analysis may appear relatively simple but the techniques and procedures applied by market analysts can be extremely sophisticated. Market studies can be developed into elaborate investigations. The levels of market analysis can be performed to reflect a spectrum of increasingly complicated methodologies. Estimates of demand are formulated differently depending on the level of analysis. In some cases, demand may simply be inferred from current market conditions, or rates of change may be used to develop projections. Because of shortcomings in this simple approach, caution must be exercised. To perform an in-depth analysis of forecast or fundamental demand with relation to a proposed property development, extensive data must be gathered and sound judgment must be applied to make projections. The analyst refines the forecast demand estimate by considering the perceptions of market participants and assessing the likelihood that current trends will continue.

Inferred analysis and fundamental analysis

The property developer can use current and historical market conditions to infer future supply and demand conditions. In addition, to forecast rates of supply, demand and absorption and then capture this over a property development's holding period, it is possible to supplement the analysis of current and historical market conditions with fundamental analysis. Table 7.3 summarises the distinctions between inferred and fundamental analysis.

Inferred analysis or trend analysis is descriptive and emphasises historical data rather than future projections. The focus can be general, with selected

Table 7.3   Types and levels of analysis

Inferred analysis A → B	Fundamental analysis C → D
Inferred subject attributes	Quantified subject attributes
Inferred locational determinants of use and marketability by macro-analysis	Quantitative and graphic analysis of location determinants of use and marketability by macro and microanalysis
Inferred demand from general economic base analysis conducted by others	Demand derived by original economic base analysis
Inferred demand by selected comparables	Forecast demand by subject-specific market segment and demographic data
Inferred supply by selected companies	Quantified supply by identifying existing and forecasting planned completion
Inferred equilibrium/highest and best use and capture conclusions	Quantified equilibrium: • highest and best use (concept plan) • timing (quantified capture forecasts)
Emphasis is on: • instinctive knowledge • historical data • judgement	Emphasis is on: • quantifiable data • forecast • judgement

Source: Fanning et al., 1994

comparable properties representing the larger market, or more specific and include area-wide market data and subject-specific conclusions. Fundamental analysis is a more detailed study of market conditions, focusing on the specific submarket of the proposed property development and providing a strong argument and quantifiable evidence for projections of future development. This level of analysis is based on the premise that real estate value is tied to the services that the completed property development provides and that a study of the market for those services will reveal influences on the value of the real estate.

7.5 Types of analysis

In addition to different levels of analysis, the discipline of market analysis comprises several related types of analysis. For a proposed property development it is important to identify which of the following types of market analysis is most appropriate:

- economic base analysis
- market studies and marketability studies
- investment analysis
- feasibility analysis.

The types of market analysis differ more in scope than in procedure. All forms of market analysis investigate local economic activity and factors influencing the supply and demand of a particular type of property or a specific market area, though not always focusing on a specific property. Also, the conclusions of these investigations all lead into the highest and best use analysis required for the proposed property development.

Economic base analysis

The economic base of a community is the economic activity that allows local businesses to generate income from markets outside the community's borders. Thus, economic base analysis is a survey of the industries and businesses that generate employment and income in a community, as well as of the functions of employment such as the rate of population growth and levels of income. Employment figures serve as a proxy for income in economic base analysis. Basic employment industries provide the economic foundation for a community by producing goods and services that can be exported to bring money into the local economy. Although some segments of the service sector can be considered basic economic activities, most service industries are not basic because the service provided and the income generated remain within the community's borders. Growth in basic employment can reflect changes in population, household income or other economic factors influencing land use and real estate value. Surveys and other data-gathering techniques employed in economic base analysis generate primary data that can be used in other types of market analysis, although limitations apply if the markets differ substantially.

Market studies and marketability studies

A macroeconomic market study provides a broad picture of supply and demand conditions for a proposed property development (e.g. office space, retail space) or for a specific area. In a market study the focus is not placed on a particular property development, although in most cases eventually a more detailed marketability study is necessary. In a marketability study the focus is placed on how a particular property development will be absorbed, sold or leased under current or anticipated market conditions. In contrast to market studies, a marketability study is property specific. It should identify the characteristics of the property's market and examine the affect on the proposed property development.

A marketability study is founded on analysis of four factors of value: utility, scarcity, desire and effective purchasing power. The interaction of these four factors will determine the marketability of a property. It should be noted that utility and scarcity are supply-side factors, while desire and effective purchasing power are demand-side factors. The development of a property usually entails both a construction (newly built, conversion or renovation) phase and a marketing phase. The marketability study must describe the demand and supply situation under current market conditions (for the estimate of 'as is' value) as well as the demand and supply situation over the planned construction period (for the value upon completion) and the marketing period (for the estimate of value upon stabilisation). In other words, a marketability study for a proposed property development must focus on each point on the development timeline for which a value is to be estimated. The demand and supply analysis must investigate market conditions, both current and future, to determine the absorption rate and other factors that will affect value during the marketing period.

A marketability study should answer the following questions:

• Who will the end users of the completed property development be? (i.e. buyers or tenants.)

- What are the characteristics of the expected end users? (Age, family size, space needs and preferences as to facilities and amenities.)
- Does the utility of the property, whether proposed or existing, satisfy the requirements of the intended market?
- What is the demand for the proposed property development that is to be marketed?
- How many end users would want the completed property development? (i.e. desire.)
- How many potential users can afford the completed property development? (i.e. effective purchasing power.)
- What share of demand is the completed property development likely to capture? (i.e. capture rate.)
- What is the supply of competitive property developments that will be marketed?
- How many competing property developments currently exist?
- How many competing property developments are under construction?
- How many competing property developments are planned?
- What is the estimated absorption rate for the proposed property development to be marketed?
- Are there alternative uses for the proposed property development that would provide a higher return on the investment?
- What are the relative risks associated with alternative uses of the property?

Discussion points

- What does a marketability study set out to achieve?
- How would a marketability study for different property types differ?

At all times, the property developer must be careful not to misinterpret data or use historical data as an absolute prediction of the future. For example, the absorption rate experienced by competitive projects is sometimes incorrectly assumed to indicate the absorption rate for the subject when it is actually an indication of demand. Consider the evaluation of a proposed office development where there are three competing office developments in the subject's market area. Over the past year, these office developments have been leasing accommodation (assuming the area per tenancy is similar) at an average rate of three lettings per month, five lettings per month and seven lettings per month. Simply using the average sales rate for the three competing office developments, five lettings per month, as the estimated absorption rate for the subject would most likely be incorrect. The total lettings for the three competing office properties can, however, be used as a broad indication of the total historic demand for similarly developed office developments in the subject's market area, i.e. fifteen lettings per month is the implied demand for this type of real estate product. The property developer should study additional market factors, including relocation patterns and the development of new competing office developments, to support the estimate of total demand over the proposed office development's marketing period.

The proposed office development's marketing period can be determined by analysing the supply of competing office accommodation in the market area, including the proposed development and all other proposed and existing office property developments. Consider the following situation:

- it is anticipated that the three existing office properties mentioned above to continue to let office accommodation during the subject's marketing period;
- another proposed office development will be added to the existing competition in the marketplace during this period;
- total demand is fifteen office tenancies per month.

Then, the average absorption rate for the five office developments will be three new lettings per month. The property developer can then determine whether the proposed property development's absorption rate will be equivalent, higher than or lower than the average rate. The reasoning for the rate chosen should be explained in the property developer's conclusion. If a marketability study prepared by another party is being used as a basis for the proposed property development, it is important to recognise this study represents secondary data. Care should be taken to carefully review the study to determine its validity and whether it can be relied upon.

Investment analysis

Investment analysis helps a property developer determine whether a specific property meets the risk and return requirements of an investor, i.e. a prospective purchaser (or owner–occupier) in the completed development. By comparing the prospective rates of return offered by alternative property developments, it is possible to estimate the required rate of return for the proposed property development. Measures of return and risk are related to standards of market performance and opportunity cost.

An investor who selects a particular property development forgoes the opportunity to invest in a different property development and will select the development that best meets specific investment objectives. Some prospective purchasers look for the highest rate of return at the lowest risk, while others seek the assurance of long-term growth at a more conservative rate of return. In addition to the illiquidity the investor endures over the term of the investment, there is a potential for opportunity cost if alternative investments at comparable levels of risk outperform the investment chosen. Interviews with prospective purchasers or owner–occupiers regarding their expectations about yield, inflation and market growth may provide support for estimated property yield rates. Actual investor projections for properties recently acquired lend further credence to estimates of property yield.

Feasibility analysis

Economic feasibility analysis is defined as an analysis undertaken to investigate whether a proposed property development will fulfil the objectives of a purchaser. The profitability of a specific real estate project is thus analysed in terms of the criteria of a specific market or purchaser. Alternatively, the term may be defined as a property development's ability to generate sufficient revenue to pay all outgoings and charges and to provide a reasonable return on, and recapture of, the initial financial outlay. Economic feasibility is indicated when the market value or gross realisation of a property development upon achievement of a stabilised condition equals or exceeds all costs of initial production. Analysing the feasibility of proposed uses requires the property developer to forecast future market conditions and the timing of events such as the sale of office accommodation in a proposed development. Inadequate analysis of development projects, large or small, can contribute to a project's failure.

Feasibility – highest and best use

Highest and best use and feasibility analysis are interrelated, but a feasibility analysis may involve data and considerations that are not directly related to highest and best use determinations. Such an analysis may be more detailed than a highest and best use analysis or have a different focus or require additional research. Generally, the options for developing real estate under a variety of alternative uses is evaluated as listed in Table 7.4. The use that maximises the completed value then represents the highest and best use.

Table 7.4 Comparison of different types of real estate studies

Goal/purpose	
General market analysis:	Identify demand for appropriate potential uses
Feasibility analysis:	Determine values of appropriate potential uses (based on data collected during market analysis e.g., residual land value, rate of return, capitalised value of overall property)
Highest and best use analysis:	Of the appropriate potential uses, determine the use that yields the maximum value
Processes/steps	
General market analysis:	Perform supply and demand analysis for appropriate potential uses
Feasibility analysis:	Calculate NOI/cash flows of appropriate potential uses and select appropriate cap rate/discount rate to form an opinion of property values
Highest and best use analysis:	Specify terms of use, timing, and market participants (e.g., user of the property, equity investor, debt investor) and compare values of appropriate potential uses
Results (data generated)	
General market analysis:	Forecasts of absorption rates and probable rents for appropriate potential uses
Feasibility analysis:	Property value of appropriate potential uses based on respective data
Highest and best use analysis:	Highest and best use of property

Traditionally, highest and best use analysis has been associated with land residual analysis, which is derived from classical economics. In a classic land residual analysis of a proposed property development, land value is attributed to the income that remains after improvement costs are compensated. Highest and best use of the land as though it were vacant indicates only how the land should be used if it were vacant. In some instances, the existing land value by itself can exceed the combined value of the land and buildings in its current use. Although highest and best use analysis is primarily a tool used when assessing the basic land value, for a completed property development it can also be used to measure a building's value contribution on the assumption that completed property development value minus land value under highest and best use equals improvement value. While buildings can be modified and changed, the essential characteristics of a site (e.g. location) cannot. The income to any particular site depends on the use to which it can be put. Land in any specific market is, therefore, a function of the income that land can generate and sustain.

Seven-step process

Most market analysis assignments can be performed using a seven step process presented in Table 7.5.

Market analysis assignments can be elaborate undertakings, particularly if a large amount of primary research is required. The following examples outline the procedures and thought processes required when using this process to analyse the markets for different property developments. In this process, an investigation of the economic overview of the market for the proposed property development normally precedes the procedures that make up the market analysis.

Office space demand

To forecast the demand for existing or additional office space in a particular location or district over a given period, an analysis is conducted of the relationship between supply and demand in the overall market area and the district's actual and potential share of the existing and projected demand. The time when a proposed property development will reach stabilised occupancy can be forecast in this way. Demand for office space in the overall market area is estimated with the following steps.

1. **Property productivity analysis**   Type of tenancy and quality of accommodation are primary identifiers of an office building's competitive status. Physical items of comparison include:

    *   building design and construction materials
    *   signage
    *   exterior lighting
    *   street layout
    *   saervices and utilities
    *   parking
    *   lot and building lines
    *   landscaping and grading
    *   office space layout
    *   tenant finish
    *   floor sizes

Table 7.5   Seven-step market analysis process

Step 1   Property productivity analysis

Identify which features of the proposed property development that affects productive capabilities and potential uses of the property. Those attributes can be physical, legal, or locational, and they will be the basis for the selection of comparable properties.

Step 2   Market delineation

Considering the potential types of property development possible, identify the market for the defined use (or more than one market if the property has alternative uses).

Step 3   Forecast demand

Economic base analysis is the basis of the analysis of

existing and anticipated market demand. Examine population and employment data to analyse and forecast demand. The scope of work required by the assignment (as well as time and budgetary constraints) will dictate to what extent demand-side variables must be investigated.

Step 4   Competitive supply analysis

Marginal demand is established through analysis of existing and anticipated supply of similar property developments

Step 5   Equilibrium or residual analysis

Investigates the interaction of supply and demand to determine if marginal demand exists and then makes predictions as to when the market will move out of equilibrium.

Step 6   Forecast subject capture

Comparing the productive attributes of the proposed property development to competing property developments, then an estimation can be made of the market share the subject is likely to capture considering market conditions, demand, and competitive supply.

Step 7   Financial modelling

Consider and model alternative development configurations with the emphasis placed on financial feasibility analysis. This will compare the productive attributes of the proposed property development to competing property developments, then an estimation can be made of the market share the subject is likely to capture considering market conditions, demand, and competitive supply.

- stairways, corridors and lifts
- electrical system
- heating, ventilation, and air conditioning
- amenities
- security
- building management and tenant mix.

Locational considerations for office buildings are often analysed both in terms of the subject's location within a cluster of office buildings and the proximity of that office submarket relative to other office submarkets in the competitive office market area.

2. **Market delineation**   In contrast to retail catchment areas, which are defined by the consumers they serve, an office market is tied more to the businesses that seek to occupy offices in that location. Usually the market area for an office building is generally spread over a broad metropolitan area, with law firms and financial institutions often seeking space in prestigious, centrally located buildings, while businesses providing other types of services may prefer suburban offices with ample parking facilities and reasonable rents.

3. **Forecast demand**   To estimate office demand, the analyst must investigate various types of information:

- size of the workforce occupying office space, segmented by occupational type;
- size of the workforce occupying office space in the subject's class;
- minimum area of office space per worker on a per square metre basis;
- normal vacancy rate for this type of office building.

Future projections may be made in annual, biannual or multiple-year increments. Note that if a 10-year forecast is being developed and steady growth is anticipated, the demand for the first period is subtracted from the demand for the last period and the difference is divided by the number of periods in the forecast to yield an annual demand estimate.

4. **Competitive supply analysis**   In addition to competitive space currently being developed or in the planning stage, the competitive supply of office space in a market may also be affected by demolitions, renovations and the conversion of space now under other uses, e.g. convert residential to office or visa versa. Information on proposed office properties may be difficult to obtain, especially reliable information on the timing of new construction and its completion. Important characteristics of competing properties include:

- size (gross floor area and net lettable area)
- age
- vacancy level
- access
- parking
- tenant quality
- building management
- building quality and condition
- amenities
- support facilities.

5. **Equilibrium or residual analysis**   The comparison of existing and projected demand for office space with the total supply of current and anticipated competitive office space should consider the potential demand for upgrading or downgrading for higher grade and lower grade buildings, i.e. some tenants upgrade from a lower grade to a higher grade space in a depressed market (when rents are falling), while others downgrade from higher grade to lower grade space in an active market (when rents are increasing). When conducting an in-depth analysis it is important to consider office space subject to pre-leasing and office space that will become vacant when current tenant leases expire. If demand for office space is anticipated to grow at a steady rate, then the total supply that is available for occupancy may be divided by projected annual demand to determine the absorption period. At the end of the absorption period, additional office space will be required. This point in time represents a 'window' for the property developer.

6. **Forecast subject capture**   To determine a particular area's share of the overall office market projection, property development patterns in the area must be analysed. Central business districts are characterised by the greatest density of property development, while suburban office complexes attract tenants with lower rents and easier access for both employees and customers. However, not all suburbs share equally in the market for office space. Development patterns in areas that closely resemble the subject district should be compared. Key demographic features such as total population, educational and income levels are believed to be closely correlated with the ability of a suburban area to support an office building. It will help to develop a ratio by dividing the amount of existing office space in the district by the amount of office space in the overall market area. However, this ratio only reflects the locality's 'fair share' of the office market and may not provide an accurate forecast. Market preferences must also be considered in determining the ratio. To forecast when a proposed office building will reach stabilised

occupancy, the next step is to estimate the construction period and an absorption rate based on pre-leasing and the historic performance of competing office buildings. Historic performance is interpreted and used to forecast expectations, but it must be considered in its proper context. Performance may have been especially high during periods of rapid growth and unusually low during periods of stagnation. Detailed data on occupancy levels may describe not only geographical patterns, but also absorption rates for different building types (e.g. low rise and high rise) or different building classes (e.g. premium, high grade, medium grade, low grade) and different occupants (e.g. anchor tenants or non-anchor tenants, corporate management, research and development departments, professional services).

7. **Financial modelling** Identify and model different office development options with the emphasis placed on financial feasibility analysis. Consideration should be given to maximising the use of site in the context of the existing demand and competing supply.

Retail space demand

To forecast the demand for an existing or proposed regional shopping centre development (or a stand-alone retail shop in a high street) at a specific site over a given period (say, 5 or 10 years), the following steps can be undertaken:

1. **Property productivity analysis** Analysis of the legal, physical and locational attributes of the proposed retail centre and competing centres in or near its trade area focuses on current industry standards. Note that retail properties can become outdated quickly as industry standards change. Particular attention must be given to the following attributes of the proposed property development site and improvements:

   - land-to-building area or plot ratio
   - building area
   - parking
   - frontage, visibility and signage
   - topography
   - services and utilities
   - landscaping
   - design and building layout
   - amenity features
   - store size
   - store depth
   - tenant mix and marketing.

Locational factors are also extremely important for retail properties. The locational attributes that should be investigated include:

- land uses and linkages with the surrounding community;
- site location in relationship to patterns of urban growth;
- proximity to competitive supply.

2. **Market delineation**   Effective analytical tools for defining the primary and secondary catchment areas of a proposed shopping centre have assisted to define the potential market. The most commonly used techniques include:

- catchment areas, in which preliminary catchment area boundaries are adjusted for the specific geographic, demographic and economic characteristics of the community;
- gravitational models, a variation of catchment area circles that takes into account the effects of competition;
- customer spotting, a more detailed form of catchment areas where actual customer addresses are surveyed to determine distances and linkages.

3. **Forecast demand**   Inferred analysis of retail demand may include study of the following:

- economic base and city growth trends;
- citywide retail centre occupancy;
- competitive centre occupancy.

Fundamental demand for retail space requires further examination of market data including:

- overall population of the catchment area;
- number of households;
- average household income;
- percentage of average household income spent on retail purchases;
- percentage of retail purchases typically made at shopping centres similar to the subject;
- percentage of purchases made at the shopping centre allocated to primary and secondary catchment areas;
- volume of sales per square metre of retail area required to support the subject;
- normal vacancy rate in the market.

The estimates of inferred and fundamental demand can be reconciled with a ratio analysis of the catchment area in which the current amount of occupied retail (per square metre per person) is compared to the future population forecast. The findings of these studies may be further adjusted to account for retail income from outside the catchment area and leakage of potential retail income to other areas.

4. **Competitive supply analysis** As for other property types, an inventory of competitive retail space covers:

  • existing competitive properties;
  • properties currently being developed;
  • planned property developments for which building permits have been obtained;
  • proposed property developments.

  To complete the analysis, the supply of competitive space is rated according to:

  • size
  • access and location
  • quality of merchandise
  • reputation
  • rental rates
  • vacancy
  • tenant mix.

  The analysis of competitive supply should yield estimates of the area (square metres) of specific competition, the market rent the subject can expect to generate in the current market and a comparative ranking of the subject.

5. **Equilibrium or residual analysis** The difference between supportable lettable retail space and the amount of existing and anticipated retail space will be the estimate of additional space needed. Sales per square metre in individual retail stores may also indicate the performance level of an existing shopping centre, the centre's share of the market and whether there is opportunity for expansion. This data may be used to check the reasonableness of the estimate of additional space demanded. If there is a current surplus of retail space, the forecast of market conditions may identify when in the future the available retail space will be absorbed and demand for additional retail space will be available.

6. **Forecast subject capture** Because retail concepts can change so quickly, subject capture is especially difficult to forecast for retail properties. In addition to inferred analysis of historical capture rates of the subject and competitive properties, several fundamental methods can be used to support an estimate of subject capture:

- quantitative ratings of the proposed property development and its competition;
- the size-of-the-centre technique, in which the drawing power of a shopping centre is related to its size relative to competing retail properties;
- ratio analysis, which is applied like the size-of-the-centre technique but segments demand to the subject property only.

7. **Financial modelling** Consider different retail development options using financial feasibility analysis. Consideration should be given to maximising the use of site and the overall property development.

Industrial property

Market analysis for a proposed industrial property development is complicated by three factors:

1. the market areas for these properties are more widely scattered;
2. demand is more limited;
3. supply is highly differentiated according to the operation of the enterprise.

The market for industrial real estate reflects the unique characteristics of the property type. For example, more expensive industrial machinery is generally custom built and, except for the flexible space in multi-tenant research and development (R&D) facilities, industrial plants are typically custom designed to the needs of the particular production line. Therefore, investors in and users of industrial real estate have usually made a long-term commitment and the property developer should be aware of this common arrangement. Many older industrial firms are precluded from ever moving due to the difficulty and expense of relocation, although newer industrial facilities are less specialised, providing more flexibility in the marketplace in anticipation of growing tenants moving to larger facilities or tenants leaving for other reasons.

Industrial plants are often built with custom financing, which is the result of lengthy negotiation. Transactions may vary considerably even for similar industrial properties, particularly when a business is sold along with the real estate as part of the transaction. In the latter situation, transfer details may be confidential so market data will not be readily available. Market analysis is generally much easier for multi-tenant warehouses and distribution centres than for facilities housing more specialised industrial operations.

1. **Property productivity analysis**   Location and access to transportation are primary determinants of a distribution facility's competitive ability. All industrial properties need access to an adequate supply of skilled labour to meet both the current demand and any anticipated future growth in the industrial sector. For example, if warehouse tenants provide parts or raw materials for manufacturing operations in the immediate area, proximity to those businesses is essential, and access to major trade routes is more important to large distribution centres that serve a wider market area, such as a regional distribution hub for a major retailer. Manufacturing plants that produce potentially hazardous waste materials need to be located near, or have affordable access to, disposal sites. Physical elements of comparison include:

   - size (plot ratio, gross floor area and net lettable area)
   - ceiling height
   - loading capacity
   - the ability to control temperature
   - proportion of office space in the proposed development
   - automated operations (including the use of robotics and other evolving technologies)
   - services and utilities
   - security
   - building management and tenant mix
   - environmental regulations

2. **Market delineation**   Established trade routes can define the boundaries of the competitive market for multi-tenanted industrial space. Because warehouses and distribution centres must be close to major transport routes or railways, industrial development will tend to cluster around those features, especially major intersections in major cities where a large percentage of the region's population can be within a relatively close drive.

3. **Forecast demand**   Demand analysis for industrial space is similar to the procedure for analysing office space, but the analysis of industrial demand must take into account the functional limits on the use of industrial property and the different physical characteristics of warehouses and distribution centres. Note that less emphasis is placed on general population change. Export activity may be a better indicator of industrial demand in a market area than population growth because the businesses that occupy warehouse space generally serve a wider clientele than the local community. The property developer will investigate the following:

- employment in manufacturing, wholesale, retail, transportation, communications or public utilities;
- cost of available labour force in relation to alternative locations;
- patterns and directions of industrial growth and development, which often cluster along major highways and around intersections;
- presence of raw materials;
- exchange capability.

For retail storage and wholesale distribution properties, the level of retail sales in a market may serve as an indicator of demand for that type of industrial space.

4. **Competitive supply analysis** Because industrial operations are such a fundamental part of a community's economic base, information on the competitive supply of warehouse space and vacancy levels is often summarised in freely available research reports. Competing properties can be compared in terms of:

- size, particularly in relation to other industrial buildings
- age
- vacancy level
- access
- building management and tenant quality
- building quality and condition.

Industrial building size and tenant quality are particularly important. Large, single tenant distribution facilities usually do not compete with smaller, multi-tenant warehouses, and a building housing several closely related industrial tenants may not compete with buildings with more diverse tenant mixes.

5. **Equilibrium or residual analysis** Industrial real estate markets can react to increasing demand with more agility than the markets for other types of properties because raw storage space is easier to construct than most other sorts of buildings with more intensive finishes. When comparing the existing and project demand for industrial space with the total supply of current and anticipated industrial space and past absorption trends, an industrial real estate market's potential for change at short notice must be acknowledged.

6. **Forecast subject capture** As long as the forecast period is not extended too far into the future, the share of marginal demand that a warehouse or distribution centre can expect to capture can be estimated with about as much certainty as the capture rate for office space. Historical absorption rates may help support an estimate of the general length of cyclical shifts

in demand and supply for industrial space of the proposed property development's type.

7. **Financial modelling** Although severely restricted usually due to agreed tenant demands, varying industrial development options can be examined using financial feasibility analysis.

Hotel demand

The source of demand for hotel rooms depends largely on the nature of the proposed development, i.e. whether it is a commercial establishment, a convention hotel, or a leisure or resort property. A proposed hotel near an established suburban office park would probably target business travellers, where the future absorption of office space in that submarket may be a good indicator of demand growth in the commercial sector. On the other hand, a large resort hotel in an undeveloped coastal area would draw from a much different demographic, and the market analysis process would differ, including the sources of data used.

1. **Property productivity analysis** In general, the following attributes of a hotel's site and proposed improvements are important factors in determining the property's competitive ability:

   - size
   - room rate structure
   - overall decor and physical appearance
   - quality of management
   - Hotel chain affiliation
   - quality and character of the market area
   - facilities and amenities offered
   - income per available room, which is a common unit of comparison used in the accommodation industry to compare the income of competing facilities.

   The importance of these factors may depend on the type of lodging being analysed. Access and visibility will be more important factors in the competitive ability of a main road property, but amenities will be more important for a resort hotel.

   The location of a hotel often indicates the likely customer:

   - airport hotels and main road hotels cater to transient guests;
   - central business district hotels in a town or city draw both tourists and business travellers;

- hotels in suburban locations often rely on adjacent commercial or industrial businesses;
- convention centre hotels or resort properties are themselves the destination rather than any nearby land use.

2. **Market delineation**   Defining the market area for a proposed hotel can be difficult because this type of property does not necessarily rely on households in nearby communities to generate demand. Instead, linkages to sources of visitations in the area can be more significant than the characteristics of the surrounding neighbourhood. The development of hotels often occurs in clusters, and the emergence of a new cluster nearby can have an impact on the competitiveness of existing properties.

3. **Forecast demand**   The inferred analysis of demand for hotel rooms may include study of:

- travel and tourism data;
- hotel employment data and convention centre activity;
- office space absorption and employment statistics related to wholesale and retail trade, financial, insurance, real estate and other such services;
- occupancy rates at competitive lodging facilities in the subject's class and market area.

Fundamental analysis of the demand for hotel rooms is based on historical occupancy and room rate data. Interviews with demand generators such as major employers or officials at chambers of commerce or visitor information centres may yield information that supports an estimate of hotel demand calculated from occupancy figures. Data useful in quantifying hotel demand includes:

- number of nights per stay
- mumber of people per room
- periods of use during the year
- prices paid for rooms
- food, beverage, entertainment and telephone usage
- methods of travel

Seasonal fluctuations in demand must be taken into account for leisure-oriented properties.

4. **Competitive supply analysis**   Information on existing hotel properties and developments under construction is generally available, but the difficulties in obtaining hotel financing and the influence of foreign investors complicate the analysis of proposed hotels. Even if market

evidence supports demand for a proposed property, new development may be hindered by external factors such as fluctuations in the economies of foreign countries whose residents invest in hotel properties. The analysis of all the hotels in the market area concludes with a comparison of the relative competitiveness of all existing and planned hotels.

5. **Equilibrium or residual analysis** The current and anticipated demand for hotel rooms, measured in total room nights per year, can be compared with the existing and planned supply of available rooms. There may be a lag between when demand is evident and when supply in the form of a new property development can be added to the marketplace to accommodate that demand.

6. **Forecast subject capture** The ratio of room nights that any hotel in a market area can be expected to capture can be derived from the fair share allotted to the property, adjusted for competitive penetration factors. The allocation of the total number of room nights demanded between competitive properties can be refined by considering customer preferences such as:

- room price
- travel distance
- quality of facilities
- amenities
- management
- image

Note that hotels with particularly high market penetration in one segment will generally have lower penetration rates in other segments.

7. **Financial modelling** Examine hotel development alternatives using feasibility analysis with the market constraints identified above.

Housing demand

Proposed residential development for single household occupation

Real estate developers often want to know how many homes they can build in a subdivision, what prices they could expect to receive for those properties and the timing of sales over an anticipated holding period. A typical market analysis for a new single-family subdivision involves the following considerations.

1. **Property productivity analysis** As in any market analysis, the first step is a preliminary analysis of the legal, physical and locational attributes

of the subject units and units in competitive subdivisions. Important characteristics of a new subdivision include:

- infrastructure
- zoning
- title restrictions
- linkages to major employers and amenities
- public planning for growth
- population trends.

2. **Market delineation**   To analyse the characteristics of likely buyers of the specified housing units, the property developer constructs a consumer profile describing income levels, household size, age and preferences. The market area of potential buyers may be defined in terms of:

- time–distance relationships, e.g. travel time to employment centres and retail areas;
- social or political boundaries, e.g. school catchments, electorates;
- manmade or natural boundaries, e.g. major thoroughfares, physical barriers;
- the location of competing housing estates.

3. **Forecast demand**   Demand for single family homes is generally analysed using demographic data. Once the market area is defined, the property analyst can compile various figures for that area:

- the current and projected population within the defined market area;
- the current and projected number of households, keeping in mind that household size varies with the age of the head of the household;
- the number of current and projected households headed by owners and those headed by renters: note that there may be an overlapping category of renters who can afford to buy.

With that population information, the analyst can break down the number of owner-headed households according to their income levels to determine the percentage of households that are or will be able to meet the mortgage payments required by local banks and lenders, interest rates and other housing costs such as outgoings for maintenance, insurance and rates. Adjusting the number of owner-occupied households that can or will be able to afford the housing by the vacancy rate in the market are measures of the existing and anticipated demand for the subject property.

4. **Competitive supply analysis**   An inventory of competitive supply includes identifying the number of:

- existing competitive properties within the subject's identified market area;
- properties under construction in that area;
- planned properties in the area for which building permits have been obtained;
- proposed properties in the area;
- maximum time for competitors to develop a new property;
- covenants and building restrictions.

The total number of competitive properties in the defined market area for the projection period can be refined by checking the total number of building permits issued against those actually put to use in recent years. In addition to quantitative measures of current and anticipated supply, this step in the analysis process includes comparison of the subject and its competition for specific amenities and attributes that give housing units a competitive advantage or disadvantage.

5. **Equilibrium or residual analysis**   Existing and potential demand can be compared with current and anticipated competitive supply to determine whether demand for additional units or square metres of housing (marginal demand) exists or when it may develop.

6. **Forecast subject capture**   The final step of the market analysis process for a proposed housing property development is to analyse the competitive rating to forecast the likely capture rate for the subject. Therefore, qualitative judgments are made regarding the relative appeal of the subject property in the marketplace that must be reconciled with the quantitative evidence of marginal demand. The goal of the market analysis for a proposed subdivision is often more than simply a forecast of subject capture. The client often also wants to know if the project is economically feasible and what prices the market will bear for the product.

7. **Model different property development scenarios**   The property developer models the feasibility of various market scenarios. The break-even point, where expected construction costs and the client's desired profit margin match the anticipated sale price, often serves as a starting point for testing pricing alternatives. The analyst can also test exceptionally optimistic or pessimistic market forecasts, providing best and worst-case scenarios.

Residential unit complex

To retain its value over time, a residential unit complex needs to be able to compete effectively in the marketplace. A property's vacancy rate may be one indicator of the relative health of a property, but market analysis for this type of property involves additional considerations.

1. **Property productivity analysis**   As for most property types, the first step in market analysis for a proposed residential unit development involves a preliminary analysis of the legal, physical and locational attributes of the proposed development and similar buildings in competitive residential unit districts. Important characteristics of an existing residential unit complex include:

   • design and appearance of the property;
   • number, size and mix of units;
   • site improvements and amenities (in individual units and for the complex as a whole);
   • parking;
   • zoning (particularly the possibility of a zoning change for potential strata title conversion);
   • infrastructure;
   • public planning for growth;
   • natural features and land-use trends;
   • linkages to major employers and amenities.

2. **Market delineation**   The market area of potential renters is similar to that of potential home buyers. The boundaries of the market area for an existing residential unit are based on:

   • time–distance relationships, i.e. the commuting time to employment centres and support facilities;
   • social or political boundaries, e.g. school districts, voting boundaries;
   • manmade or natural boundaries, e.g. major thoroughfares, physical barriers;
   • the location of competing housing developments.

   In addition, the property developer investigates the tenant profile (e.g. occupational profile, income level and other demographic information) of the proposed and the market area in this step of the market analysis process.

3. **Forecast demand**   The demand for an existing residential unit complex is forecast using both inferred and fundamental methods. The inferred (trend) analysis of the subject's market area includes investigation of:

- general growth trends
- residential construction trends
- historical absorption figures
- real rental rates.

Relevant information gathered in the fundamental analysis of residential unit demand includes:

- the current and projected population within the defined market area;
- the current and projected number of households (i.e. divide total population by average household size);
- the number of current and projected households headed by owners and those headed by renters;
- the number of households that are or will be able to meet the monthly rent on units in the subject property.

An adjustment for frictional vacancy in the market may need to be made for the proposed property development, but for existing projects the analysis usually focuses on the ability of the subject property to capture actual occupancy and, therefore, an adjustment is not necessary. Additional adjustments may be needed for upgrade demand, which is generated by the upward mobility of lower-income households and latent (or pent-up) demand, which is often a result of under-building or high financing costs that restrict new property developments.

4. **Competitive supply analysis**   The competitive supply of residential units in a market area takes into account:

- existing competitive properties;
- properties under construction;
- planned properties for which building permits have been obtaine;
- proposed properties.

To complete the analysis of competitive supply, the location, age and amenities of the subject are compared to those of the competitive properties.

5. **Equilibrium or residual analysis**   A net excess or shortage of residential unit developments in the market can be determined by comparing the results of the analyses in steps 3 and 4.

6. **Forecast subject capture** The inferred analysis of the market area is revisited along with additional fundamental analysis to generate a subject capture rate. The subject's current occupancy can be compared to the estimated number of occupied units in the market, or a pro rata share can be calculated by dividing the total number of units in the subject with the total number in the market. In addition, competitive ratings for the subject property and competitive properties can be compared. If more than one form of fundamental analysis is used to calculate a capture rate, the separate conclusions should be reconciled.

7. **Model different scenarios for developing residential units** Consider and model different configuration models and the feasibility of different market scenarios.

Reflective summary

This chapter has demonstrated that market research is of the utmost importance when undertaking a property development. If there are gaps in the knowledge base, then there is a substantially increased risk of undertaking a property development that does not reach it highest and best potential. It is important to undertake market segmentation to identify the market for the proposed property development. Attention must be paid to the varying levels and drivers of demand. For office development, this will be businesses seeking office space. Retail development will be dependent upon the location and characteristics of nearby households. An analysis of the competing supply of similar land use will ascertain the total amount of stock available to the market, as well as the relevance of equilibrium.

There are different types of market analysis possible including economic base analysis, marketability studies, investment analysis and feasibility analysis. A seven-step market analysis process has been outlined for a proposed property development, although the approach differs for office space, retail space, industrial property, hotel, housing and residential complexes.

8

Promotion and selling

8.1 Introduction

An integral component of the property development process is selling or leasing the completed project and reaching the predicted sale price or rental level as originally planned prior to construction. This can only be achieved by adopting a sound marketing plan incorporating proven and reliable promotion and selling strategies. This part of the development process should not merely be an afterthought that occurs at the end of the process, but rather be viewed as a priority that will adversely affect the success of every property development. For example, many financiers now require a pre-commitment prior to confirming the loan so marketing often commences well before the construction process.

At times, property developers are unfamiliar with marketing strategies and partly or completely underestimate the costs associated with implementing a successful marketing plan. Nonetheless, it should be a primary goal of the property developer to achieve 'full market value' for their finished product, which occurs (according to the definition of market value) when the purchaser/lessee is 'knowledgeable' about the marketplace. Since there is no centralised marketplace, as opposed to other competing investments such as the stock market, there is even higher emphasis required for promotion and selling of a property development in order to 'inform' the demand side of the property market (Forlee, 2006).

Over time the options available for promotion and selling have changing dramatically as we entered the technological age and then progressed into the twenty-first century (Klein *et al.*, 2004). For many purchasers or lessees the internet has become an important medium that can not be ignored by a successful property developer (Dixon *et al.*, 2005), and at the other same time other advertising mediums (e.g. the newspaper media) have decreased in popularity in some areas. This chapter discusses the methods and options available for marketing the completed promotion, with the emphasis placed on the promotion and selling aspects.

8.2 Methods of promotion

Regardless of the type of development, it is critical that the developer is market-orientated and fully conversant with all aspects of the market. This includes identifying and defining the target market, such as by geographical location (e.g. for a shopping centre development) or by prospective investor size (e.g. such as for a major office building) and then actively promoting to this sector. The underlying aim is to promote the development to potential purchasers and tenants. At the same time the developer should establish a realistic budget to promote a building and be prepared to increase it if required. Thus, the size of the budget will depend on the nature of the scheme and the state of the occupier market prevailing at the time. It should be noted that successful marketing is not an exact science and, therefore, the proportion of the development budget will vary between projects. In other words, some developments will practically 'sell themselves' whilst others will need a concentrated and, at times, expensive marketing campaign. The alternative option, being ill-informed about the marketplace, should be avoided where possible.

The amount of finances that a developer commits to promotion is often based on their judgement formed through past experiences. Developers who specialise in a particular type of development or who concentrate their development activities in a particular geographical area will tend to have established personal contacts and a good reputation with potential occupiers – in these instances the amount of promotion (and finances) required will be minimal. Often this type of developer will promote the project by personal contact either directly or through appointed agents. However, in the case of a large-scale development, promotional campaigns tend to be very extensive and may have to extend over considerable periods of time. Associated costs are also expensive.

Regardless of the size of the promotional budget, it is essential for the various methods used in the campaign to be closely monitored. Note that research into the source of enquiries will help to gauge the effectiveness of the impact of the promotion on potential occupiers. It is important to obtain maximum value for money and a careful analysis of all enquiries in terms of both numbers and 'quality' will help to identify the most effective methods of promotion. Careful recording and analysis of results will provide a useful database for the planning and direction of future promotional activities.

It is essential for the initial planning phase of the promotional campaign to be carried out in consultation with the team who will be involved in selling/letting the scheme, whether they are appointed agents or an in-house team employed as part of the development project. In addition, the team may include an advertising agency and a public relations consultant who are experienced in the promotion of property and especially a new property development. Prior to commencing promotional activity, the developer, agent and any consultants must identify the relevant strengths and weaknesses of the particular property, in the context of the competition, so that the advantages may be highlighted. It is critical that the developer defines the target market for the property during the evaluation and design process, through their own knowledge and experience and, where necessary, via market research. Partly due to the lack of a centralised marketplace the property market is very fragmented – for example, occupiers vary from small businesses to major companies with their own retained agents or internal property departments, and they also have different needs and perceptions. Therefore, any promotion must be specifically targeted to be cost-effective.

Discussion point

Why is promotion and selling as important as any other part of the property development process?

An early decision in the property development process is the naming of the particular building or development. It should be noted that the individual identity/name of the building, together with an associated design/logo, will last throughout the promotional campaign and distinguish the development from the competition. This identifies the building or development, which is particularly important in relation to a large shopping centre development as the name will endure beyond the letting of the individual shop units, and will be used in promoting the centre to the shopping public. With an office building the name will tend to change from time to time when let to a large, individual firm who may name it after their company name. Also, names of buildings may reflect their former historical significance or their particular location. It is important that the developer, before eventually deciding on a name, checks with the local authority and the Post Office that the name is acceptable to them, so that it may be used in the postal address. Current approaches to promoting a property are listed below.

Site boards and site hoardings

This remains one of the oldest and most cost-effective means of marketing a property. The erection of a site board is usually one of the first methods of promotion, which can be arranged as soon as the developer has acquired the site. Overall, site boards can be very effective and provide an inexpensive way of communicating the existence of a scheme even before there is any physical sign of land clearing or construction. Multiple site boards may be erected during the course of the development depending on the information available at any particular time. For example, before the final details of the scheme are known, a simple 'all enquiries' board may be erected providing the name of the developer or an agent and their contact details. Note that the standard procedure is to include communication details other than just a phone number, such as email, website and mobile numbers. For larger developments there are usually contact details for several agents listed. A site board may encourage preliminary interest but must be replaced at the earliest opportunity by a board providing information on the nature and size of the scheme. Practically all development sites will have a board giving details of the development, both during the course of construction and from the time the building has been completed until sales or lettings have taken place.

Importantly, attention should be given to the positioning of the board and its actual wording (both spelling and grammar) to ensure maximum advertisement value. For example, if the site is in a prominent location (e.g. a high street or major road) then it may be the primary method of promotion and may generate the most enquiries. The site board should be clean, lit by lighting (i.e. so it is visible at night and in dull weather conditions) and regularly maintained. A site board in a poor condition gives a poor first impression and reflects badly upon the quality of the development itself. Note that care should be taken to locate the board away from trees, which may partly or fully obscure the board when they grow. Also, the developer should ensure that the boards of contractors and the professional team are kept to a minimum and do not distract from the site board. Furthermore, if the developer sells advertising space on the site hoardings to advertising agencies then care must be taken to ensure that their positioning does not distract from the main site board.

At all times the site board is actually an advertisement and in many jurisdictions requires planning permission prior to its erection. In many instances planning permission is not a large hurdle since it is a matter that can often be dealt with under a planning officer's delegated powers. Note that some planning authorities may impose size restrictions and may dictate the maximum site board size and actual colours.

Above all, the message on the board should be simple and uncluttered with relevant information. The board should inform the passer-by at a glance of the following information:

- the name of the scheme (incorporating any logo);
- the type and size (or range of sizes) of the accommodation available, whether the accommodation is 'to let' or 'for sale';
- the name and contact details (business phone, mobile phone, email address) of the contact who can provide further information, either the developer or the appointed agents;
- the website for the property (where applicable);
- the radio frequency for information about the property (where applicable).

Space permitting whilst ensuring the site board is not too 'busy', additional information, although not always essential, may include a completion date and brief specification points highlighting the existence of, for example, sustainability features and internal finishes. Other options are also possible, such as a coloured flash containing the message placed diagonally across one of the top corners of the board.

A trend in some areas is to use decorative site perimeter hoarding to surround the development site during construction in order to promote the scheme. Decorative site hoardings have two advantages: first, they distract from the unsightly chaos of building sites, providing an attractive outlook to passers-by and, second, they can provide information about the scheme in an imaginative and informative manner to help the awareness of the development. Often passers-by will recall a property development with decorative hoardings rather than the usual contractor's painted plywood effort. At the same time, decorative site hoardings can be costly when compared with ordinary site hoardings and they also lose the benefit to the developer of revenue from advertising hoardings. In the case of a refurbishment, where protection of the building during construction is usually in the form of mesh hung from the scaffolding, it is possible to reproduce a life size image (or even a promotional message) of the proposed building onto the protective mesh.

Particulars and brochures

In today's competitive marketing environment practically every property development is supported by particulars and/or brochures of the scheme providing the potential occupier with further information in response to their enquiries. Importantly, these particulars need to be aimed at potential occupiers and agents. The nature of the particulars will depend upon the nature of the property and the target market identified by market research. The fundamental question that needs to be asked is: 'What information will the reader require in order to capture their interest, and convert it into action via arranging to view the property?' The content and nature of the particulars and brochures, therefore, need to be targeted specifically at the likely readership. For example, for a new shopping centre development the target market may be residents visiting an older shopping centre in an adjoining area. For a new office building, the target market may be existing tenants in other buildings currently leasing a in range of different areas.

The information about the available individual components in the development, usually for smaller projects, is set out on a single sheet of paper with a photograph or on a double-sided, printed, colour glossy sheet. For larger schemes, more detailed information is required and, therefore, particulars need to take the form of glossy colour brochures or booklets.

Basic particulars, which may be supported by photographs or illustrations, should contain the following:

- a description of the location of the property;
- a description of the accommodation, giving dimensions and a brief specification;
- a description of the services supplied to the property such as gas and electricity;
- the nature of the interest that is being sold if it is freehold, the nature of any leases or restrictions to which it is subject and, if leasehold, the length and terms of the lease.

Importantly, the particulars should give the name, address and contact details of the agent and developer, and the name of the person in the agent's office or on the developer's staff who is dealing with the property, so that a direct contact can be made by the potential occupier and more information can be supplied as requested. Care and attention must be taken to ensure that all information given in the particulars is accurate and not 'false or misleading' and, consequently, all information needs to be carefully checked to ensure that the particulars are accurate as far as can be reasonably ascertained. For example, this rules out the use of inaccurate or subjective language in the description of the property. Legislation varies between different jurisdictions but, nevertheless, have the common goal of protecting a prospective purchaser. This legislation also applies to any information supplied or to statements made by agents and developers in promoting the property and, therefore, applies to practically all forms of promotion.

Usually it is necessary to prepare a colour brochure describing the property, giving more details than the brief particulars referred to in a one-page flyer. As part of the marketing campaign, the brochure is sent to the people who have registered an interest and replied to any advertising, site boards, mail shots or preliminary particulars, or it may be sent directly to people who are known to be interested. Accordingly, it is important that the brochure is integrated with the rest of the promotional activities. The appearance to a potential purchaser is of great importance as it reflects the quality of the development that it describes. Most developers or their agents employ a specialist outside designer or agency with experience in property brochures since they can provide professional designers and copywriters, ensuring the most effective use of words, typography, colour and shape. Furthermore, an external designer is usually more cost-effective considering the time delay that often occurs between property developments. Most importantly, first impressions count and the brochure is usually the first introduction to a building for a prospective occupier. Therefore, it is important to demand and receive extremely high standards of design and production.

Where possible, the brochure should include maps, plans, photographs and a brief specification. All photographs need to be taken professionally and many photographers specialise in property. The brochure should provide clear directions and maps to enable the reader to find the property easily. It also needs to provide information about the locality, which will be helpful to a potential occupier. For example, in promoting a particular office location information should be provided on drive times to nearby motorways/roads, towns, airports and cities; transport routes and major cities/towns nearby; and the availability of public transport services. Also, particularly if the office is in a provincial town and the likely occupier is a firm decentralising from a major city, information should be provided on local housing, shopping facilities, availability of skilled labour and so on.

For an extensive promotional campaign it may be necessary to produce more than one brochure at different stages in the development. At times, substantial financial resources can be spent on brochures; in addition to being expensive to produce, it should be remembered they are usually quickly out of date. At first, simple particulars or what is known as a 'flyer brochure' or 'flyer' (i.e. a single-sheet brochure) may be produced showing an artist's perspective of the proposed building, and including very basic details. These will need to be replaced as soon as possible with photographs of the completed building and additional details as soon as they become available. At times, it may be worthwhile producing a tenant's guide to the building for prospective occupiers who show continued interest in the building. The guide should provide a detailed description of the building and specification, together with information on operation, maintenance and energy-saving data. Due to the relatively high expense associated with glossy brochures, care should be taken not to hand out detailed brochures to prospective purchasers/lessees who are only showing a mild interest. The underlying aim should be to provide basic information (i.e. a flyer) to parties on their initial enquiry, and then to forward a detailed brochure to parties who show a genuine interest.

For most development sites a dedicated web address is standard policy, and it is often accompanied by a three-dimensional 'walk through' or a 'fly buy'. On particularly large and complex development schemes there may be a computer simulation and/or graphics combining video and voice in a multimedia presentation. The internet profile can also be hosted on a real estate agent's site rather than a stand-alone site for the property development, although accompanying a reduced cost is the loss of identity for the development amongst all of the other competing properties on the agent's site.

For interested potential purchasers it is possible to show three-dimensional images of the building proposed if the architect is using a CAD (computer-aided design) architectural design system to produce drawings of the scheme. This technology enables the viewer of the computer screen to be effectively taken round the building from every angle and perspective. Photographic stills of the three-dimensional images produced can be also be used in any promotional material.

Advertisements

As a long-established method of raising the profile of a new development and attracting interest, advertisements are still considered a cost-effective medium. Usually the advertisements are either aimed directly at potential occupiers or indirectly through their agents depending on the target market. Such advertisements are usually placed in the property sections of local and, where appropriate, national newspapers; in the various property journals, such as *Estates Gazette, Property Weekly* and *Estates Times;* or, where appropriate, in trade or industry journals, e.g. *RICS Business.* If the target market is likely to be local businesses then clearly local newspapers are the most obvious place for advertisements. If the potential occupier is likely to be represented by an agent, then the property press is the most effective place for advertisements.

The content of an advertisement is important, but unquestionably its design and layout are crucial. It is essential that a balance is reached between too much detail (i.e. too 'busy') or, alternatively, being too vague, but the goal of the advertisement (rather than simply 'telling' not 'selling') should ultimately be to encourage action. Many advertising companies seek to accomplish the following steps, commonly referred to as AIDA, from the perspective of a prospective purchaser:

1. Attention – the advertisement catches the initial eye of the purchaser;
2. Interest – the purchaser reads the advertisement and seeks further information;
3. Desire – there is a need to gain more information on the development;
4. Action – the purchaser contacts the developer/marketer/real estate agent and seeks more information.

Good design and layout increase the impact of the advertisement, whereas poor design not only wastes money but can sometimes create an adverse impression. Thus, poor design should be avoided at all costs. It must be remembered, particularly in the property press, that the advertisement will compete with others and, therefore, must be designed to make an impression. The cost will vary with the size and position in the paper or magazine. A front page advertisement, while costly, may provide the opportunity to create a bold, imaginative advert that will make a lasting impression. The cost of a one-off advertisement on the front page has to be weighed up against the cost of a series of smaller advertisements in less prestigious positions. The style and advertising media should be varied in a constant attempt to improve results.

Care needs to be taken over the way in which the advertisement is worded and presented and in nearly all cases it is worth employing an experienced advertising agency to ensure the maximum impact, although some large firms of agents have an in-house advertising department. The advertisement should contain sufficient information to attract the potential occupier, but it should not be so cluttered that it is difficult to read. Information should include:

- the type of property, e.g. industrial;
- approximate size building area (and land area where applicable);
- services available, e.g. electricity, lift;

- location and proximity to transport and amenities, e.g. a train station for an office building;
- if the property is for sale or to let;
- the telephone number/email address of the agent or the developer, or both, should be given, so that the potential occupier can make a direct contact.

At times it may be difficult to judge the effectiveness and worth of an advertising campaign, but some attempt should be made to identify the source of any enquiry. Of course, it must be remembered that some readers might see an advert and take no immediate action, but the information will stick in their memory and when they are subsequently looking for accommodation, they will recall the advertisement and make enquiries. This is referred to as 'top of the mind' advertising and used by billboard advertisers on the side of motorways.

Newspaper and magazine advertising is a common form of advertising where costs can vary depending on the readership and the competition by other advertisers for space in that medium. Other factors, such as the actual page number or photos in black and white versus colour, will adversely affect the overall advertising costs.

Relatively little use is usually made of television and radio for individual properties, primarily due to cost restrictions. Other negatives including the 'one-off' nature of an advertisement (i.e. either you see/hear or miss it) and the limited audience (i.e. off-peak periods are cheaper with a smaller audience). Some developers argue that radio and TV are only used when all else has failed, although advertisements on local commercial radio during 'drive time' are often effective to spark interest for certain projects.

At times, poster advertising is a medium that can be used to promote larger properties, employing prominent locations such as railway stations, roadside hoardings and even on or within public transport and taxis. As an example of its effective use, consider the promotion of an out-of-town office development with ample car parking. Posters displayed in appropriate railway stations, in the underground, on buses or in taxis can be an effective and often overlooked means of advertising to those who are forced to travel in crowded, uncomfortable conditions into the capital. A poster advertisement, like a site board, should convey its message instantaneously as people will only have time to glance at it and rarely to study it carefully. However, poster advertising is expensive and is difficult to target specifically although it should always highlight the web address access to detailed information.

Internet presence

As the property market is part of the information society it is important to appreciate the wider role of stakeholders (i.e. people, government and business) with regards to information and communications technology (ICT) (Dixon *et al.*, 2005) – see Figure 8.1. From an ICT perspective a successful property development must have a presence on the internet in some form or another. Many developments have their own domain name or web address, which provides detailed up-to-date information about the project and saves the cost of printing expensive colour brochures. Other sites have live webcam feed of construction on the site so interest is maintained. In addition, many newspapers/television/radio advertisements highlight a web address that prospective purchasers/lessees can instantly access for more information and contact details. At the same time the developer can use a counter and monitor the number of times that each web page is accessed, especially after a newspaper/television/radio advertisement is run – this has been proven as a reliable means of accurately measuring the cost-effectiveness of the original advertisement.

It is essential that a professional web page designer is engaged to design and maintain the web page. Although practically anyone can construct a simple web page in Microsoft Word (e.g. in .html format), a professionally designed web page is well worth the extra cost (Klein *et al.*, 2004). As well, there are other benefits such as the time taken to load the website (note

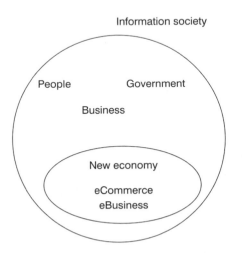

Figure 8.1   The information society and the new economy (Source: Dixon *et al.*, 2005)

that prospective purchasers/lessees using the internet are used to fast loading web pages and will not wait more than a few seconds), the use of multiple links to other pages and photos/videos, the use of sound (where applicable) and an impressive window to the world. It is recommended that examples of impressive web pages are studied and noted through the course of daily business to suggest as examples to the web designer. Finally, the costs outlaid for web design and maintenance must be in proportion to the size of the property development and in the context of the overall advertising budget. There are not hard and fast allocation budget models, but generally a project with a broad global purchaser base would be reached more efficiently via the internet, as opposed to advertising in multiple newspapers in other regions/countries.

Discussion point

How has technology impacted upon property development from a marketing perspective?

Mail shots

Direct mail shots may be effective and relatively cost-effective. Mail shots are used on their own or to support an advertising campaign. However, because it has become such a widespread promotional method great care has to be taken if it is to be really effective. They should be aimed at a very carefully selected list of potential occupiers and the success of the shot will depend very much on the compilation of the mailing list, and any follow-up procedures. There are a number of specialist direct mail organisations capable of producing mailing lists with a degree of specialisation and accuracy. They maintain general lists of industrial or commercial firms that can then be broken down into particular categories such as company size, trade and location. There is a limit to the frequency with which direct mail shots can be used and the employment of one of the specialist firms will often be found to be advantageous. Care must be taken in selecting firms: they must routinely and frequently update their mailing lists. To be effective the mail shot must reach the right person within the organisation, who is responsible for property matters, and at the right address. To overcome the almost automatic tendency of the recipient to throw the contents of the direct mail shot into the wastepaper basket, make sure that the message can be seen at a glance. A very long letter that has to be read through to the end before the message is fully understood, or a brochure or leaflet without any covering message, will often fail. By contrast, a short, sharp covering message attached to a brochure or leaflet will often be absorbed by the recipient before they have time to throw it away. The letters should be prepared in such a way that each one appears to have an original signature.

With mail shots, unlike advertising, the results and effectiveness of any promotion can be quantified. It can be also followed up by a telephone call to obtain a reaction or information on general property requirements. Such telephone canvassing may be carried out by the agents or a representative employed in the show office/suite. Mail shots, provided that the target market is accurately identified, can be cost-effective.

Email

In this age of technology email has become one of the primary communication mediums. Email has many benefits including low delivery cost, low maintenance/upkeep costs and the ability to send out bulk emails. As many prospective purchasers/lessees do not have the time to answer the phone (they are at work, at home or elsewhere), for many email has emerged as a viable and convenient alternative. To send out the same information via postal mail is comparatively expensive and has an accompanying time delay, and there may be the occasional incorrect or old address. Detailed information can be accessed via the email using links (instead of email attachments), rather than sending out emails that have a large size.

For larger projects it is possible to subscribe to regular information emails from the developers that keep the reader up to date, as well as allowing the developer to gauge the interest in the project. In addition, it is possible to email special offers or the date of the final completion. Be wary, a balance must be struck between saturating the reader with too many emails and not keeping them informed; the core aim should be to encourage action by having them visit the property development and/or commit to a detailed discussion. There is a danger that the emails could disseminate too much information and, therefore, not give the recipient any reason to seek more.

Launching ceremonies

For larger property developments (e.g. a major office building) it may be appropriate to launch the development by conducting a ceremony during the course of construction or by an opening ceremony when the building is completed. It is normal to invite to such functions the local and national press, local and national agents, and sometimes potential occupiers. Local councillors or their officers may also be invited to help with the development company's public relations. For example, the opening of a shopping centre may be quite a grand affair with a celebrity to perform the opening ceremony. Entertainment should be imaginative with a definite theme, perhaps extending from fancy dress to the type of food. It is important to pay great attention to detail; a ceremony enjoyed by those who attend it will be remembered for a long time.

Whatever form of ceremony or party is decided upon, it is of paramount importance that at the time of the function the development is in good order. At an opening ceremony the buildings must be ready for occupation with all services in working order. In the case of a shopping centre, the major units may be let and occupied, and the opening of the development will be linked to the opening of those units.

In an increasingly competitive environment developers have had to make their ceremony more imaginative and innovative in order to encourage agents to attend. In other words, agents will receive many invitations to ceremonies and they have to make a choice about which to attend. Accordingly, such incentives as free gifts, a prize draw to win a holiday or a night in a major hotel may have to be offered. The main advantage of this type of promotion is that it brings agents, particularly those retained by a potential occupier, to the property and gains/retains their attention.

Show suites and offices

A promotional campaign must be timed to have its maximum impact when the building is completed, so that photographs can be included in the material and potential occupiers can view the final product. Due to the nature of property development and time required prior to completion, the developer must be creative in marketing the final product. Thus, it is essential that the developer places importance on selling/letting the building before completion to maximise the cash flow. During the construction period any promotional activity can only be supported by the use of plans, artist's perspectives, models and CAD images to help build up a picture of what the completed project will look like. Accordingly, particularly on large, mixed-use or phased developments (e.g. a business park), it is worthwhile to set up a show office on site during the course of construction to display plans, models and promotional material of the scheme.

The office, often a temporary structure that has been appropriately decorated, painted and fitted out inside, should be staffed by an on site representative of the developer or agent who is able to talk knowledgeably to agents and prospective occupiers who arrive on site. Refreshment facilities should be available to offer to the prospective purchaser/tenant. A 'finishes' board should be on display showing samples of the internal finishes proposed for the scheme. It is important to make sure that the show suite or office is clearly signposted and, if possible, the landscaping should be brought forward to enhance the environment. In any event, any landscaping in a scheme should be planted as soon as possible, allowing for planting seasons, so that it can mature slightly before completion of the whole scheme.

Also, when the development has been completed, its appearance is of vital importance. In the case of an office scheme or an industrial scheme with high office content, it will often be advisable to fit out a floor, or a part of a floor, as a show suite with partitions and furniture, so that potential occupiers can see clearly how the offices will look when they are occupied. This may be supported by plans showing alternative open plan or cellular layouts. It may also be advisable to fit out the main entrance hall and reception area, and it is important to ensure that the common parts and lavatories are kept clean. In nearly all cases, carpeting will be provided by the developer throughout the building, and if lettings prove difficult, it may be that this will be extended by the developer offering to fit out or furnish an office for a particular tenant. In the case of a shopping scheme or warehouse scheme, where units are in shell form, a sales office on site is useful to welcome potential occupiers.

Public relations

Predominantly larger developers (particularly public quoted companies and REITs) either retain public relation (PR) consultants or have in-house staff responsible for relations with both the general public and the press. Everyone responsible for public relations should be involved in a particular development scheme right from the beginning and be kept informed at all times on the progress of the scheme. Their main priority is to promote and enhance the reputation of the development company as a corporate entity, whilst there will be spin-offs in relation to the promotion of individual development schemes. Where a scheme is particularly sensitive in relation to local politicians and the local community it is worthwhile to employ a PR consultant specifically for that scheme to ensure it all runs smoothly.

Through their press contacts PR consultants can achieve useful press editorial at significant stages in the development programme. For example, carefully controlled and timed press releases may be made to the property press on any of the following occasions: initial acquisition of the site, obtaining planning permission or consent, the completion of the scheme, and on the letting of all or part of the scheme. Editorial coverage is often offered by a publication if advertising space is being booked to coincide with coverage of a particular geographical area or a particular type of property. At times, editorial coverage is much more likely to be read than an advertisement and can be achieved at no cost.

Promotional material should be on display at every public relations event where either the developer or the development is being promoted, such as corporate hospitality events and exhibitions.

Discussion point

List the various types of promotion and the positive and negative aspects of each option.

8.3 The role of the agent

A real estate agent is usually responsible for selling/letting a particular property on behalf of a developer. Although some developers have an experienced in-house selling team, it is not always possible to rely solely on their efforts. The appointment of an agent will depend upon the circumstances of a particular scheme and the likely market. In addition, many agents offer certain advantages over an in-house agent.

As a starting point, agents are often in a better location for attracting business. For example, for an out-of-town developer undertaking an industrial development in a provincial town it makes more sense to employ an agent with offices in the particular provincial town. A good firm of agents should have detailed knowledge of the particular market in which they operate, being thoroughly familiar with current and future levels of demand and supply. They may also have additional specialised knowledge and experience, such as specialising in certain types of property. Furthermore, they will also have personal contacts with agents retained by a specific client or decision makers within the property department of a potential occupier. In order to be effective, the agent must be continuously involved in the marketplace so that they are aware of changes in market conditions.

Agents can be retained on a 'sole agency' or 'joint agency' basis. As sole agent they alone are responsible for disposing of the property and are entitled to a fee on each letting or sale. Often a normal range of fees might be 10 per cent of the first year's rent for a letting (allowances will be made to take account of any rent frees or inducements given to the tenant on a letting) or 2 per cent of the sale price for a sale, although this can be negotiated depending on the size of the development and the relationship between the developer and the agent. A joint agency arises where two or more agents are instructed to sell the same property. This can often happen where both a national and local firm of agents are instructed together. In such a case, on each letting or sale, the developer will have to pay a larger fee, perhaps one and half times the normal amount, with the agents sharing the fee between themselves on some agreed basis. For example, joint agents would receive 7.5 per cent each from an agreed fee of 15 per cent of the first year's rent for a letting instruction. The appointment of a second or third agent (if two are already appointed) is often not the decision of the developer, but a condition of a funding agreement, where the fund wishes to see their own agents involved.

Often a developer can employ a national or regional firm of agents operating with or without the help of a local firm. There is no general rule to apply; each case must be looked at on its own merits, but a national agent normally has a greater understanding of the larger and more complex schemes and has more direct and frequent contacts with the larger companies and multiple retailers which are attracted to them. Also, they can offer a wider service to include investment advice if the scheme is to be forward-funded or sold on completion. On the other hand, the local agent will often have a better understanding of the particular characteristics of the local market and of local occupiers and retailers.

Regardless of that type of agent or combination of agents is used, it is important that they can have the opportunity of contributing to the planning, design and evaluation of the project. They should be able to draw attention to features of the design that add to or detract from the marketability of the property. Also, they must be able to comment on prices or levels of rent, the nature of competitive development and the most effective time for letting or selling. It is recommended that they are engaged at an early stage, which will allow them to become thoroughly familiar with the property they are going to sell or let. In other words, it is annoying to a potential purchaser or tenant than to find that the selling agent cannot provide full details of the property that they are offering. In addition, they should attend regular meetings with the client to be kept informed of progress on the development and to plan the promotional activities.

Most importantly, it is essential to distinguish between the appointment of agents and offers to pay commission to an agent introducing purchasers or tenants. There is a contractual relationship with the former, the terms of which (both expressed and implied) need careful consideration. The agency agreement should make clear the length of time the appointment will remain in existence and how it may be terminated, specifying any retainer payable and setting out the rate of commission and in what circumstances and when it will be payable. In addition, it will specify whether the agreement is for a sole or joint agency and what the developer's rights are concerning their discretion to employ additional agents. Also, it should be made clear whether agents are entitled to expenses, whether or not they succeed in disposing of the property.

At times, offers will be made to pay commission to agents on the signing of legal agreements for sales or lettings for purchasers or tenants introduced by the agent. No formal agency appointment is made in these circumstances, and this type of arrangement is substantially different from the formal appointment of agents to act for the developer. Sometimes the offer is open to anyone introducing a purchaser or tenant, sometimes the offer is made to a limited number of agents. Although a developer might think that an offer to pay introductory commission to almost anyone would be likely to result in the most business, this is not necessarily so. For example, if too many agents are handling the property there is a danger that it will be 'peddled around the market' and this may create an unfavourable impression on the ground. In other words, after an initial flood of enthusiasm many agents tend to lose interest if they know that the property is in a large number of hands, and some of the best agents might be reluctant to be involved with properties widely and indiscriminately offered in such a way. The smarter agents will devote their resources to potential sales/lettings where they have the highest likelihood of success, i.e. not shared with a large body of other agents. Sole or joint agents, or perhaps three or four working on an agency basis, are much more likely to have a sense of involvement. Whatever the arrangements, it is important to keep all the agents up to date at all times.

Communication and record keeping is critical, so that it can be seen at a glance which prospective purchasers or tenants have been introduced by which agents and precisely when. The question of disclosure is important. Note that if an agent/s has already been retained to find accommodation for the prospective purchaser or tenant, they will not be able to accept an introductory commission from the developer. Furthermore, when agents refer to 'clients', they are normally retained and will not seek a commission; when they refer to 'applicants', they will normally expect a commission because they are not retained by the applicants. In all cases, the position should be clarified explicitly at the outset.

When an agent is appointed by a developer they should be involved at the commencement of the development process. One of the agents appointed may have even been responsible for initiating the development process by introducing the site to the developer and/or arranging the necessary development finance. Often the agent/s has valuable input, ranging from their knowledge of both the occupier and investment market through to the evaluation and design processes. Developers should be aware that agents will always tend to be cautious when advising on design and the specification of the scheme from a letting/selling point of view, since they wish to ensure that the scheme has the widest tenant/purchaser appeal. Developers should make their own judgements of the advice received, supported by their own experience and research. More importantly, both developers and agents should obtain feedback from tenants on similar schemes.

Regular meetings and communication with agents is essential and should normally take place on a monthly or fortnightly basis, largely depending on the stage reached in the development process. At these meetings the agent should report progress in relation to the letting/selling of the scheme. The report will normally take the form of a schedule detailing any enquiries received from interested parties and any known requirements for accommodation that are to be followed up. The agent should advise on the availability of competing buildings and what terms are being quoted on those buildings, and concluded sale or letting transactions on similar accommodation should be reported. The source of each enquiry should be noted so that the developer can judge the effectiveness of a particular method of promotion, e.g. internet, newspaper advertisement. These promotional activities will also be discussed and the agent should have a full input in any decision on the content and design of all promotional material. At times the agent will be responsible for the booking and scheduling of all advertisement space, unless an advertising agency has been appointed. As both agents and advertising agencies book space on a regular basis they should be able to pass on the benefits of any discounts and should be able to obtain good positions within the relevant publications.

8.4 Sales/lettings

After a potential occupier or an applicant has expressed interest in viewing the property, then the agent will show them round in the first instance. If continued interest is shown, then the developer will tend to be involved in all future viewings and negotiations with the interested party. However, there is no substitute for personal contact with the applicant, although there may be times, particularly early on in the negotiations, where the agent should lead the discussion. The developer should make a judgement as to when it is appropriate to become involved.

During their regular meetings agents will be able to advise the developer when it is appropriate to quote terms (for sale/let) and at what level of price/rent. Agents should know the flexibility they have in negotiating terms, bearing in mind the requirements of the financier or the investment market. In fact, the terms of any letting will be determined by both the method of funding the scheme and the intentions of the developer as to whether it will be retained as an investment or sold to an investor. If a scheme has been forward-funded then the fund will need to approve the terms of any letting and the type of tenant (refer to Chapter 4). If the intention of the developer is to sell the completed and let scheme to an investor, then the terms of any letting must be acceptable to the investment market at which it is aimed. As we have already discussed in Chapter 4, this will depend on if the property is considered to be 'prime' in terms of both its quality and location. The developer should be advised by the agent who will be responsible for handling the investment sale. Furthermore, the developer must continually remember that the terms of any letting will directly affect the investment value of the completed development. Even though a property may be initially retained by the developer, thought must be given to any eventual sale in structuring the terms of a lease so it is generally acceptable to investors.

Now it is important to focus on the various terms that need to be agreed when negotiating with a potential tenant. Note that if negotiations are being held with an owner–occupier then the only terms to be agreed are the price and any work to be carried out by the developer.

The demise

The accommodation being let needs to be defined and this is commonly known as the 'demise'. A decision needs to be made, in the case of an office development, as to whether lettings of single floors or part floors are acceptable to the financier and/or the developer. In many instances, to maintain the maximum investment value, agents will be instructed to seek a single letting of a building. A multi-let building will narrow the potential investment market due to the additional management and risk involved, which will be reflected in any investor's required yield. Immediately following each downturn in the property cycle there are always many examples of schemes that remain vacant, and often this is complicated when a developer has strictly adhered to a policy of accepting only single lettings. However, a decision as to whether to relax such a policy and accept multiple lettings will need to be made if the letting of the building proves difficult. Other factors come into play here, and the level of flexibility will depend on the financial status of the developer and/or the influence of any financier or development partner. One question needs to be asked: 'Is it better to maintain cash flow at the expense of maximising the capital value?' Or, alternatively: 'Is it better to have a building 100 per cent let at a lower rate per square meter or 50 per cent let at a higher rate per square metre?'

If the demise is not self-contained, then consideration needs to be given to adapting the premises to suit other tenants. For example, rights may need to be reserved for the tenant in respect of the use of common parts, such as staircases, lifts and toilets. Arrangements will also need to be made in respect of separating services for metering purposes and heating/air conditioning.

Rent

The developer will usually quote an advertised rental value for the accommodation in consultation with the agent, based on an assessment of the open market value for the property. The developer will also calculate the rent required, which when capitalised at the appropriate market-derived yield will provide the developer with a satisfactory return. This return should compensate the developer for their exposure to risk. At the same time, it is important for the developer and the agent to be aware of all recent letting transactions and the rents being quoted on other properties, in order to be familiar with the interaction of supply and demand. The real estate agent should be able to assist with interpreting the level of demand via the number of enquires for the development. The level of supply is also critical, which would be quantified by examining the number of buildings under construction or planned, as well as recent sales in the area to developers. Note that when analysing letting deals and quoting rents, allowances must be made for any differences in the specification and location of the comparable properties.

The actual rent agreed between the developer and a potential tenant will depend on how the negotiation is conducted and the willingness of each party (Reed, 2007). Also, it will be affected by the strength of their respective bargaining positions in the prevailing market conditions. For example, in difficult market conditions where supply exceeds demand and tenants have considerable choice, the developer may consider offering inducements in order to maintain the required rental value for the property. However, such inducements should be kept within reasonable limits otherwise the developer will be trading inducements in return for a rent in excess of a market rent: the developer will be 'over-renting' the building in the view of any potential investor. At times, reasonable inducements, such as 6 months rent-free (to allow time for the tenant to fit-out the building), or a capital contribution towards specified fitting-out works (e.g. fit-out) are considered in many markets to be an accepted practice by potential investors. A fitting-out contribution will usually be related to fixtures and fittings that are referred to as 'landlord's fixtures and fittings' for rent review purposes. Therefore, when the rent is reviewed in accordance with the terms of the lease, account will be taken of the benefit of those landlord's fixtures and fittings in determining the rent for that particular property.

In the past, during periods of severe oversupply in the property market, there have examples of substantial inducements or incentives (e.g. from 12 months up to 3-year, rent-free periods) being offered in order to maintain a certain rental level on office schemes. Importantly, this would ensure the building is 100 per cent let at market rates, although the inducement may not be so transparent to a potential purchaser of the building. In other words, this would appear to maximise the investment value for the developer especially if the rent review clause in the lease is drafted on the basis of an upward-only rent and, in the process, ignoring the existence of inducements being offered in the open market at the time of the rent review. A downside of achieving a higher rent is that in a period of low rental growth, potential investors will discount the value of the slice of the rental income considered to be in excess of market rent as they will see no growth until after the first review.

Lease term

The circumstances surrounding the length of a lease and the existence of any lease options for many property developments will depend largely on whether the property is viewed as a potential investment for an institutional purchaser. In other words, the strength and value of the lease (or income flow) will be linked to the risk, if the development is to be relied upon as a reliable source of money in the future. Most institutional investors prefer long-term leases to decrease the risk of a tenant void and loss of income. Over time the length of the lease term has been changing to suit tenant demands – shorter leases have been emerging as tenants have been in a better position to negotiate such terms to suit their business plans. They have also been able to negotiate break clauses at certain specified times within their leases to allow flexibility as their business changes. Some institutions may, depending on the circumstances, accept the existence of such clauses provided they agree to the lease term as well. Developers may also negotiate a landlord's break clause to coincide with a redevelopment or refurbishment opportunity.

With forward-funded arrangements the institution will dictate the length of the lease in the funding agreement and any flexibility will need to be strongly argued by the developer in the light of market conditions. For schemes where the likely tenants are small businesses or sole traders with little or no financial track record, such as in a specialist shopping scheme or small industrial units, then the developer will often be more flexible due to the limited market demand. Such schemes will not be considered 'prime' due to the type of tenant and, therefore, institutionally acceptable lease terms are not so important.

The tenant

The final value of the development will be linked to its ability to earn money and the strength (i.e. reliability and security) of the tenant. The financial status and background of all potential tenants will need to be checked by the developer to ensure that they have the ability to pay not only the agreed rent but also all outgoings on the property. Usually the institutional investors prefer that the tenant's profits exceed a sum three times the rent. In most cases the developer will need to see at least the last three years' accounts of the tenant to establish the tenant's financial standing. In the case of new businesses or those without sufficient track record, then the developer should obtain bank references and trade references. In addition, the business plan and cash flow projections should be examined to show the long-term projections over the same time frame as the proposed lease.

If the developer or fund consider the financial covenant of a tenant is insufficient and, therefore, too risky, then bank and/or parent company guarantees may be sought to reduce perceived risk and provide some form of guarantee in the event of non-payment of rent during the lease period. Alternatively, with private companies and sole traders, directors' personal guarantees will be sought.

On a retail scheme the developer and/or the fund may wish to influence the tenant mix within the scheme to ensure a variety of retail uses for the shopping public, who will in turn determine the success of the scheme. A diversification of different tenants will ensure the shopping centre can cater to a wide variety of shopping needs, which may alter as seasons change or shopper demands vary. Also, having tenancies of different areas (square metres) will be more attractive to a tenant than if there was only one size available.

Repairing obligations

The responsibility for repairs and maintenance during the term of the lease will depend largely on whether the demise is self-contained. For example, in the case of a single building the sole tenant will be responsible for all internal and external repairs. On the other hand, where the demise is only part of a building that is shared with other tenants and/or the owner, then that tenant will usually be responsible for internal repairs and the landlord will be responsible for the external and common parts. At times it will depend on how well each party (i.e. the landlord and the tenant) negotiated the original tenancy agreement and what was agreed. In order to recover overall costs, the landlord's expenses in repairing, servicing and maintaining the common parts will usually be recovered by a service charge directly levied on all tenants, most often in proportion to the area of their demise. Also, developers must provide prospective tenants with estimates of the service charge.

Other considerations

There are many other factors to be considered in the process of promoting and selling/letting a completed property development. The information listed in this chapter gives a broad overview of the basic principles, although the circumstances will vary much in the same way that each property itself differs. Nonetheless, there is by no means an exhaustive list of the matters to be considered by the developer when negotiating with tenants.

In order to be binding, every letting or sales transaction must be legally documented in the form of a lease or a contract and transfer of title. It is important that the developer instructs a solicitor as soon as the promotional campaign begins so that all the necessary draft documentation can be prepared. Once negotiations are concluded with a tenant or purchaser the legal work can then proceed quickly. A developer will also need to ensure all the necessary draft deeds of collateral warranty have been agreed with the professional team and contractors (with design responsibility), should either the tenant or the purchaser wish to benefit from them. A tenant is likely to require warranties on schemes where the repairing obligations could be considerable if something goes wrong and they wish to be able to pursue the relevant professional or contractor for a remedy. In such an example the remedies under a warranty are limited and a tenant may wish to benefit from decennial insurance to protect against latent defects. However, the developer will need to judge the likelihood of such insurance being required by either a tenant or purchaser as the option to arrange cover must be arranged before construction starts.

Prospective tenants and purchasers will typically make enquiries through their solicitors as to the existence of services and will want to see copies of all planning and statutory consents (Adams *et al.*, 2005). Any tenant or purchaser will also require full information on the building in the form of 'as-built' plans and maintenance manuals. The developer should, therefore, ensure that all the necessary supporting paperwork is in place before draft contracts and leases are issued, to reduce the risk of delays in the legal process. Overall the developer should be prepared to answer all queries and supply any documentation related to the development.

Most importantly, the development process does not end the once the documentation has been completed and the keys are handed over to the tenant or purchaser. The developer will have ongoing responsibilities such as ensuring the satisfactory completion of the defects identified at practical completion and assisting the tenant with any 'teething' problems. The developer will have a continuing relationship with the tenant as their landlord unless the property is sold either for occupation or as an investment. Even if there is no continuing contractual relationship, the developer should maintain contact with all their tenants or occupiers and provide an after-sale service. It will pay future dividends and help the reputation of the developer in the market, as failure to protect a reputation can adversely affect future proposals, although the developer may never find out the reason. For example, occupiers may not speak to each other but they may talk to their agents if they have cause to complain about a particular development. In today's economic environment a tenant/purchaser will be much more vocal when things are going wrong that when things are going very well. Thus there is no substitute for direct feedback from which invaluable lessons can be learnt for future developments. Overall there is a strong argument that the development industry as a whole should research occupiers' needs to a much greater degree than is currently undertaken. At all times the occupier is the customer and they should come first.

Reflective summary

In order to reduce exposure to risk, the task of securing of an occupier should be at the forefront of a developer's plans at the start of the development process. During the evaluation stage it is critical to research and accurately quantify the level of occupier demand for the development in the proposed location, and where further research should be carried out to identify the target occupier market and exact requirements in terms of the design and specification. At the same time consideration should be given the level of competitive supply in the market and when it will be available. However, such research is often overlooked by developers, which may result in a building being difficult to dispose of – this can not be solved by simply throwing money at promotion. The aim of promotion is to make potential occupiers aware of the development scheme and for a campaign to be effective it must be carefully targeted at and tailored to its audience. Agents, with their knowledge of the market, have an important role to play in disposing of the development to an occupier, whether through a letting or sale, together with any subsequent investment sale. Any letting should be secured on terms acceptable to the investor market to maximise its value and reduce perceived tenant and property risk.

Sustainable development

9.1 Introduction

This chapter establishes what is meant by the terms sustainability and sustainable development. The concept of corporate social responsibility (CSR) is explained and the relationship between sustainability and CSR is described. Why property is an important sector in relation to sustainability and CSR and identification of the key areas of impact are illustrated. How these impacts are being addressed by the UK government and by developers, investors and occupiers is also described. Finally, an illustrative case study demonstrates how sustainability and CSR are adopted in practice.

9.2 What is sustainability and sustainable development?

There could be as many definitions of sustainability and sustainable development as there are groups trying to define it. To a large extent all the definitions are concerned with:

- comprehension of the relationships between economy, environment and society;
- equitable distribution of resources and opportunities;
- living within limits.

To some degree, different ways of defining sustainability are useful for different situations and different purposes and, for this reason, various groups have created definitions. The definition most frequently referred to is the one established by the United Nations World Commission on Environment and Development report, 'Our Common Future', in 1987. This definition states that:

> Sustainable development is development that meets the needs of the present without compromising the ability of future generations to meet their own needs.

This landmark definition, also referred to as the Bruntland definition after the chair of the UN Commission, established two key principles that have influenced the sustainability debate significantly. The principles are those of inter-generational and intra-generational equity. Inter-generational equity is the principle of equity between people alive today and future, as yet unborn, generations. The inference is that unsustainable production and consumption today will diminish and degrade the environmental, social and economic basis for tomorrow's society, whereas sustainability involves ensuring that future generations will have the means to achieve a quality of life equal to or better than today's. Intra-generational equity, on the other hand, is the principle of equity between different groups of people alive today and is about social justice issues, therefore it manifests itself in areas such as alleviating the debt and improving health of developing countries.

Another important or influential definition is that of the World Business Council on Sustainable Development (WBCSD, 2000) who state that:

> Sustainable development involves the simultaneous pursuit of economic prosperity, environmental quality and social equity. Companies aiming for sustainability need to perform not against a single, financial bottom line but against the triple bottom line.

As with the Bruntland definition above, the WBCSD definition covers social, economic and environmental aspects. Significantly, there is reference to the triple bottom line: a concept introduced by John Elkington in 1994, which challenges the traditional approach to accounting in business. Triple bottom line (TBL) accounting means expanding the traditional reporting framework to account for environmental and social performance as well as economic performance. Consequently, TBL requires that a company's responsibility is towards its 'stakeholders' and not solely its shareholders. Significantly, the term stakeholder refers to those influenced, either directly or indirectly, by the company's activities. This concept has been articulated by O'Riordan *et al.* in Figure 9.1.

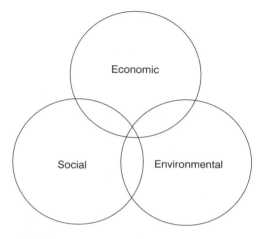

Figure 9.1   'Three Pillars' model of sustainable development (Source: O'Riordan et *al.*, 2001 cited in RICS, 2007)

The argument is that all three aspects have to be taken into account for sustainable development. Over time, as social values change, these ideas have evolved into the concept of corporate responsibility (CR) or corporate social responsibility (CSR).

9.3 Corporate social responsibility

What is CSR? Many definitions exist, some of which are more helpful than others. For example, the European Union (EU) glossary defines CSR as companies who have 'been encouraged to develop socially and environmentally aware practices and policies' and this definition covers part of what CSR is about. A more detailed definition is that CSR is a company's obligation to be accountable to all stakeholders in all its operations and activities, with the aim of achieving sustainable development not only in the economical dimension but also in the social and environmental dimensions. This definition incorporates the social, economic and environmental aspects in the WBCSD definition above. The WBCSD use the following definition: CSR is, 'the continuing commitment by business to behave ethically and contribute to economic development while improving the quality of life of the workforce and their families as well as of the local community and society at large'.

In essence, CSR is about taking a broader approach to business activity. Historically, businesses were motivated solely by financial or economic gain and paid little or, in some cases, no regard to environmental issues such as the pollution that resulted from their activities or social issues such as exploitation of labour in developing countries. Businesses that adopt CSR are committed to following best practices and to making all their activities open and transparent.

The roots of CSR can be traced back to the philanthropy of Victorian industrialists, such as Joseph Rowntree, Titus Salt and George Peabody and, in the 1920s, American industrialists Andrew Carnegie and Henry Ford. CSR re-emerged in the late 1970s and early 1980s as organisations became increasingly concerned about their public image (Clark, 2000). Since then CSR has mainly been driven by large companies, many of which voluntarily issue CSR disclosures in their annual reports and on their websites (Williams and Pei, 1999; Lewis and Unerman, 1999).

Although CSR has become entrenched in the business community, there is still debate about what CSR means. Some see it as acting ethically (Esrock and Leichty, 1998), others as pragmatic public relations (Frankental, 2001). The former view is based on the belief that organisations have a moral obligation 'to pursue those policies, to make those decisions, or to follow those lines of actions which are desirable in terms of the objectives and values of society' (Esrock and Leichty, 1998). The latter is based on the idea that CSR is an attempt at gaining competitive economic advantage by creating a favourable impression on employees, clients and consumers (Burke and Logsdon, 1996). In between these two extremes is the notion of CSR as the obligations of a company to its stakeholders, that is 'the people and groups who can affect or who are affected by corporate policies and practices' (Lantos, 2001). This is the stance taken by most companies at present. A 2003 survey by PricewaterhouseCoopers (PWC) revealed that the majority of companies stated that being socially responsible involved:

- providing a healthy and safe working environment for employees;
- creating value for shareholders;
- supporting community projects;
- good environmental performance.

However, the survey revealed that with CSR one size does not fit all. There were significant differences across sectors and from country to country. It is necessary to understand that there are different ideas of what CSR should mean in different countries. For example, in the USA CSR is perceived more as a form of philanthropy, where businesses make profits, unhindered, and then they donate to charitable causes. It is seen as a blemish for the business to get benefit from donations. In many African countries CSR is seen as the link between corporate profitability and social improvements. In Europe and North America, CSR is viewed as environmental responsibility, ethical economics and charitable giving (PWC, 2003). However, the European concept, which the UK adheres to, is targeted on operating the core business in a socially responsible way, complemented by investment in communities for solid business case reasons. Some argue that this approach is more sustainable as CSR becomes an integral component of wealth creation, adds to business competition and increases the value of wealth creation to society. Furthermore, in difficult periods there is reason to exercise CSR further, whereas if CSR is philanthropic outside the core business mission and objectives, it will be jettisoned when times are hard.

Currently, most of the information disclosed by businesses in their voluntary social disclosures relates to employees, the environment and the community (Milne and Chan, 1999). Some studies have examined companies' CSR disclosure practices (Esrock and Leichty, 1998; Williams and Pei, 1999), which showed that:

- There are variations in the way that companies approach CSR. For example, some companies have teams or individuals dedicated to ensuring that the company fulfils its obligations, whereas others use CSR as a form of public relations.
- There are differences in the amount and type of information disclosed. For instance, some provide detailed information of past performance in areas such as health and safety, environment and community relations, where others present the CSR strategy.

This suggests that the drivers for CSR vary across sectors and between companies, that perceptions of CSR differ and that a true indicator of the real value that companies attach to CSR is where, if at all, they locate this function within their organisational structure (Frankental, 2001). For most, the primary driver behind CSR is increased profitability. Studies in this area have primarily used one of three methods to measure CSR: expert evaluations; content analysis of voluntary social disclosures and other corporate documents; and analysis of performance in controlling pollution. The research has shown that companies' financial performance goes some way to explaining variations in approaches to CSR and disclosure practices. For example, philanthropic activities were seen to be affected by profits and sales, and a combination of high CSR performance and high levels of disclosure had a positive impact on companies overall profitability. The underlying argument is that CSR can pay off, although it depends on how individual companies approach the issue and a variety of other factors. For example, companies exist in many different guises and degrees of complexity, ranging from those that are profit orientated in the private sector to those in the public sector that are not profit orientated. The activities in which companies are engaged varies considerably and it is not surprising that there are differences in CSR.

9.4 The different types of CSR

There are a number of types of CSR (Lantos, 2001) and these types have evolved from economic, legal, ethical and philanthropic responsibilities and include Lantos's three types of classification – ethical, altruistic and strategic CSR. According to Lantos (2001) ethical CSR is the moral, mandatory fulfilment of a corporation's economic, legal and ethical responsibilities, even if the business might not appear to benefit. Altruistic CSR equates to philanthropic CSR and involves contribution to the good of society, even if it compromises profits. Altruistic CSR might be the funding of schools or donations to drug and alcohol programmes. The justification is that corporations, through their wealth, have the power to affect parties beyond participants in its transactions. Some argue that altruistic CSR is not legitimate for corporations because it involuntarily involves shareholders (through lower share prices), consumers (through higher prices) and workers (through lower pay). Reasons against company involvement in CSR are that firms are not competent to involve themselves in public welfare issues and that, through taxation, corporations already contribute to society.

However, there is wide consensus that strategic CSR is a legitimate activity for corporations. Strategic CSR is defined as caring, corporate, community services that accomplish strategic business goals. Corporations contribute not only because it is morally right but also because it is in their best financial interests to do so, hence fulfilling their duties to stakeholders (Friedman and Miles, 2001). Within strategic CSR the voluntary aspect leads to increased employee morale and higher productivity, attracts customers to a caring corporation, or contributes to the local community making it easier to attract good employees. Overall, strategic CSR presents a win-win situation benefiting the community and the firm.

However, for strategic CSR to operate effectively, corporations have to be able to measure and benchmark their activities. Without measurement, CSR is reduced to public relations and little else. Strategic CSR has to embrace the stakeholders, to be rewarded by financial markets, to be defined in relation to goals of ecological and social sustainability, to be benchmarked (PWC, 2003), to be audited and to be open to public scrutiny. Finally, it has to be embedded vertically and horizontally within the business and embedded in the corporate planning function. Five criteria: centrality, specificity, proactivity, voluntarism and visibility are requisites for strategic CSR and examples of CSR behaviour relating to them are identified (Burke and Logsdon, 1996).

Furthermore, it is noted that a company's approach to CSR is additionally complicated and affected by its legal position. Companies are legal entities with certain rights and privileges, but also certain liabilities. Another factor that may affect a company's approach to CSR is its position and level of power within its sector. For example, some companies have a turnover that exceeds the gross national product of some small countries, and their power and influence can be immense. Indeed, 'big business' is different and these corporations may have a different perception of CSR as a result. Therefore, given the variation in conceptual understanding and the different types of CSR, UK companies will incorporate CSR in a variety of ways. Furthermore, for the property developer there is likely to be some variation within and across the sectors and according to company size.

9.5 Sustainability reporting

At the heart of CSR is accountability to all stakeholders for a business's social, economic and environmental impacts, and sustainability reporting is the method of demonstrating accountability. Reporting also enables business to develop new targets and goals in terms of its social, economic and environmental impacts, thus enabling performance gains to be made. Over time protocols have been developed for businesses to adopt for sustainability reporting. Initially this might involve stakeholder consultation to identify the goals and target relevant to the business. It is imperative to benchmark activities, for example, the amount of waste products produced by a company or the amount of water or electricity consumed per annum might be measured. After benchmarks are made, targets can be established such as reducing electricity consumption in the company's offices by 10 per cent per annum. It is necessary to set up a strategy whereby reporting and documentation of the policy can be recorded within set time frames. Equally, a responsible person has to be identified within a senior management role so that they have the power to ensure targets are implemented within the business. The number of people engaged in CSR depends on the size of the business and its activities. An annual report is prepared by the CSR manager and made available to the public, usually on the company website. Within the annual report new targets will be set out of the forthcoming period.

The Global Reporting Initiative (GRI) was set up in 1997 to provide business with a reporting framework to ensure consistency across and within business sectors with reference to sustainability reporting. The GRI is a non-profit coalition of over fifty investor, environmental, religious, labour and social justice groups. GRI elevated sustainability reporting to parity with financial reporting by determining metrics applicable to all businesses, as well as sets of sector-specific metrics and a uniform format for reporting information relating to a company's sustainability performance. The GRI Sustainability Reporting Guidelines recommend specific information related to environmental, social and economic performance and is structured around a CEO statement, key environmental, social and economic indicators, descriptions of relevant policies and management systems, stakeholder relationships, and management, operational and product performance as well as a sustainability overview. For further information see http://www.globalreporting.org/Home.

Closely affiliated with CSR and GRI is ISO 14001, *Environmental Management Systems* (EMS). This was established in 1996 to set out a framework for businesses to manage their business in a more environmentally aware manner. For example, ISO 14001 established a system of record keeping, auditing, reporting and managing a business to identify environmental impacts and to set targets for reductions. Such reporting has to be publicly available and is typically found on company websites in an annual CSR or Environmental Report. Companies are required to identify key personnel within their organisation who have responsibility for managing the EMS. There are now a whole series of ISO 14000, international standards on environmental management, providing a framework for the development of an environmental management system and the supporting audit programme.

Discussion points

- What is sustainable development and how does it relate to property development?
- What are the key aspects of CSR and how does it relate to property development?

9.6 Property development, sustainability and CSR

There are a number of ways that CSR and property development interface. Given that CSR is about the quality of a businesses management (in terms of people and processes) and the nature of, and quantity of, their impact on society, CSR may affect property development either in the direct employment of staff and how they operate and occupy their property or in the way the developer develops property; that is by developing sustainable property.

To put into context why the property sector is so tied in with sustainability, buildings impact on the environment in many different ways. For example, buildings are responsible for 40–50 per cent of all energy end use. Climate change and global warming have been linked to manmade emissions of greenhouse gases into the atmosphere. Carbon dioxide is a greenhouse gas and is given off during the consumption of fossil fuels such as oil and gas. Typically, electricity has also been generated through gas and coal power generation plants, therefore electricity consumption also releases carbon dioxide into the atmosphere. Some fossil fuels have higher carbon content, for example black coal has a lower carbon content than brown coal and thus emissions will be lower when black coal is used to generate electricity.

Furthermore, the operation phase of a building's life cycle is crucial in its overall environmental impact; buildings are responsible for 40 per cent of the total waste going to landfill sites. Buildings also consume 16 per cent of fresh water and 40 per cent of raw materials. Finally, 25 per cent of global timber harvest is related to buildings. These statistics show clearly that the overall environmental impact of buildings is substantial and needs to be reduced. This state of affairs has been acknowledged for some time by the UK government, the EU and also research organisations such as the Building Research Establishment (BRE). The key areas of impact are described in more detail in 'Key areas of impact in property development and sustainability' below.

9.7 Arguments for sustainability

The environmental, social and economic arguments for sustainability in the built environment are clear. Environmentally, the Working Group II for the Fourth Assessment Report on climate change for the Inter-Governmental Panel on Climate Change (IPCC) concluded in 2007 that there is a high degree of confidence in concluding that humankind's activity has lead to changes in the planet's climate (IPCC, 2007).

The 2007 report followed three earlier reports that have tracked changes reported in research worldwide since the 1970s. With each report more evidence has accumulated and the degree of confidence stated in the conclusions has increased. Such reports have influenced many countries to adopt internationally agreed targets with regards to greenhouse gas emissions, such as the United Nations Framework Convention on Climate Change Kyoto Protocol signed in 1997 (UNFCCC, 2007). The targets that each nation commits to varies depending on their economic development – for instance, EU signatory nations agreed to limit their greenhouse gas emissions to 1990 levels between the periods 2008 and 2012. Over 160 countries have signed the protocol agreement and there is acceptance of the environmental arguments for sustainability, though there are some exceptions such as the USA, although at state (e.g. California) and local government levels there are moves to reduce emissions. The EU regards climate change seriously and is adopting various strategies to reduce their environmental impact. The EU also has a significant influence on environmental legislation across all member states.

Of the social and environmental arguments posited, sustainable buildings are promoted as being healthier buildings for occupants and users – for example, because natural materials are used in the specification, which leads to less off-gassing of volatile organic compounds (VOCs). Furthermore, sustainable buildings tend to maximise natural daylighting and fresh air thus improving both the internal air quality and the internal environment quality in their design, which research has shown is a preference for building users.

There are claims that sustainable buildings lead to less absenteeism from workers, less churn or turnover of staff and increased productivity in workers, which creates an economic argument for sustainable buildings. Another aspect of the economic argument is that sustainable buildings cost less to run. Similarly, if less waste is produced by building users and more waste is recycled, the running costs are reduced. Water consumption can be substantially lowered in sustainable building design allowing users or owners to enjoy lower water bills. However, it should be noted that some claim the initial costs of sustainable buildings are higher than non-sustainable buildings and this may be the case depending on the range of sustainable measures incorporated into the design. On the other hand, not all sustainable buildings are excessively more expensive and it is often possible to make some savings elsewhere in the specification to offset the cost of the sustainability features. Finally, once the property market recognises and fully accounts for the added value in sustainable buildings in market valuations, the issue of high capital costs will diminish. There is an argument that sustainable buildings will attract higher rentals because of the perceived health and productivity gains for occupants attributed to them.

9.8 Key areas of impact in property development and sustainability

Clearly property development has some very significant impacts on sustainability issues. This section of the chapter outlines the different stages of the property development cycle covered in the previous chapters and identifies the sustainability issues that affect that stage.

Development land and sustainability

The sustainability issues that affect land include loss of habitat and bio-diversity, and contamination of land either by natural causes or as a result of a previous use. Protection of natural resources is high on the government's agenda. Development of land with contamination can add significantly to costs and developers need to take appropriate steps to reduce their risks when acquiring land that has been previously used.

Development finance and sustainability

As well as the traditional sources of finance for property development, new areas are emerging as financial institutions and lenders adopt and promote CSR and risk management strategies. These developing forms of finance either make sustainability a requirement of the finance package and/or offer incentives and discounts on finance for sustainable developments. Eco-finance is a developing area as financial institutions integrate sustainability into their policies, practices, products and services. There is an awareness of the potential benefits for banks in integrating sustainability into their business strategy. In 2005, 86 per cent of 120 financial institutions reported positive changes as a result of integrating social and environmental issues in their business.

The UK-based HSBC bank has adopted environment-related policies and procedures. The list includes guidelines on dangerous chemicals, freshwater, infrastructure and forest products. Underpinning HSBC's diminishing appetite for environmentally sensitive transactions is its environmental risk standard, launched in 2002, which minimises the environmental, credit and goodwill risk associated with the bank's investments. HSBC's due-diligence register, features environmental impact assessments and reviews by external auditors. As banks globalise, they have to consider the broader context of sustainability.

Other lenders screen environmental risks surrounding corporate loans to help clients improve their regulatory compliance and environmental management programmes. As environmental risks become more complex the financial community is thinking outside its traditional approach. For example, impacts may occur over longer timescales and across borders. To reduce the exposure of commercial loans, banks place importance on businesses' ability to manage environmental liabilities. Environmental consultants estimate the nature and likelihood of risks and their advice informs the bank's decisions on whether to accept, avoid, manage or mitigate risks, or to seek insurance cover. This works when risks are quantifiable and there is certainty; however, the qualitative nature of many risks generates ambiguity. Consultant panels are used on contaminated land work where the consultants recommend actions to minimise risk and liability and may be made a condition of lending.

Sectors and industries designated as having high environmental risk include waste management, forestry, and oil and gas, which sometimes involve developers. The environment-related policies and procedures, adopted by the UK-based HSBC Bank, outline the environmental and social impacts of lending, covering risks, regulation and international best practice. The banks identify clients' environmental risks and help to reduce exposure. Where necessary, loans are made conditional on clients taking measures to reduce risks. Loan decisions are informed by three risk considerations: direct, dominated by land contamination; indirect, including regulatory impacts and changes; and reputational. Thus, larger development schemes may be considered by the banks under these provisions as well as smaller property development schemes that are on land previously used or contaminated. The trend is that these provisions will be extended over time by the number of banks using them and also the scope of the considerations the banks take into account.

Planning and sustainability

The key sustainability issues that relate to planning are:

- transport
- ecology and site issues
- zoning and land-use issues.

Transport impacts on work, leisure and recreation patterns and on the environment in which we live. The increasing dependence on cars and road freight has significant environmental, social and economic impacts. Carbon emissions from transport accounts for a quarter of total carbon emissions for a developed nation. The social impacts revolve around the frequency and severity of accidents, the impacts on health from inhaling emissions, whilst the economic impacts revolve around the costs of social and environmental impacts. For a sustainable environment we need to provide access in a way that has less impact. In planning terms this can be achieved by developing and implementing policies that:

- improve and promote walking, cycling and public transport and changing habits to reduce car use;
- manage freight transport by moving more by rail and reducing heavy truck traffic;
- make streets, bus stops and tram stops safer, including lower traffic speeds and better security;
- reduce oil dependence and shifting to cleaner, renewable energy for transport;
- ensuring transport impacts are reflected in investment decisions and the costs that users pay;
- plan in a more integrated way to involve the community and link land use and transport.

When property development schemes are considered, the typical transport issues include:

- Access to public transport nodes and facilities. For example, environmental assessment schemes advocate that a proportion of a development has to be within a certain distance of 30-minute peak and hourly off-peak public transport services. The Building Research Establishment Environmental Assessment Method (BREEAM) Residential requires that 80 per cent of a development is within 1,000 metres of such services as a minimum for urban schemes.
- Provision of bicycling facilities and changing rooms for cyclists. Assessment schemes state minimum requirements expressed as a percentage of number of dwellings or occupants depending on the property type.
- Proximity to local amenities such as banks, shops, pharmacy, school, medical centre, place of worship, children play area and so on.

Ecology and site issues centre on the loss that results from developing land and destroying or impacting on the local ecosystems such as flora and fauna. There is merit in selecting land that has little ecological value and areas of high ecological value are becoming difficult to develop. There have been high-profile media stories of developments that have impacted on ecology such as the extension to the M3 motorway in Hampshire, which, in 1994, attracted 5,000 people in protest. The site was occupied causing delay and additional costs to the developers. Security firms were employed to keep activists out. Where ecological features exist on a site, protection should the goal of the development team. Developers are able to enhance the ecological value of sites and proposals are welcomed by planning authorities. Therefore, developers must consider the ecology and site issues at an early stage to avoid negative attention from environmental groups or unwanted media coverage. Ecology issues can be benchmarked by measuring the ecological footprint of the building.

Sustainability on a regional and local level can be affected by which land is zoned for different uses. Clearly, the property developer has less influence individually on the regional plans. However, they should identify the prevailing trends in the locations or regions in which they operate. Consultation with the local authority plans will highlight authorities' intentions with respect to zoning and land-use issues.

Sustainable design and construction

There are numerous ways the construction phase of property development can be environmentally friendly. Firstly, there is the selection of the contractors on the tender list, secondly, the procurement process and, finally, the activities during construction itself.

Some contractors have adopted CSR (see 'Corporate social responsibility') and are committed to reducing environmental impact and adopting social responsibility in their business operations. They have a vision statement on their website setting out their goals and drivers in terms of environmental, social and economic sustainability factors, such as the example in Appendix 1 from the Bovis Lend Lease website. These publicly available statements allow property developers to select contractors on the basis of environmental credentials. Furthermore, the websites outline progress towards sustainability visions and goals. It is necessary to ensure that the organisations have set targets that they monitor and audit. An annual sustainability report should detail relevant information regarding sustainability. Frequently, case studies of sustainable building projects illustrate the contractor's expertise in these types of development. Other ways of selecting contractors would be on past performance in respect of the construction of sustainable buildings. Developers can also build up a list of contractors with whom they work well on certain projects and this list may form the basis of selection for a project.

Design and build issues will depend on the property type and location. Key sustainability areas that impact on property development are:

- reducing carbon dioxide emissions;
- minimising pollution;

- Life Cycle Costing (LCC) or Whole Life Costing (WLC);
- using resources efficiency.

Approximately 52 per cent of the UK's carbon dioxide emissions come from constructing or using buildings. Energy efficiency is a way of reducing emissions. In the UK, providing more insulation, more efficient glazing, and introducing measures such as recovery of heat from waste water or air and individual meters for heating and hot water reduces the energy consumed and leads to cost savings. Such measures also improve indoor air quality.

Embodied energy is the energy that it takes to create the materials used in construction. Materials such as brick or concrete have a high embodied energy because the manufacturing process has a very high energy demand. However, offset against this must be the capacity for a material to retain heat, which will then be released back into a building. It is necessary to work out what is the best combination of materials for a project. The intended use has to be taken into account, along with the heating system and other installations to be used, not just the construction technique and materials.

Renewable energy has an important role in reducing carbon dioxide emissions. There are many renewable technologies that can be used, some are more appropriate than others for particular projects. Renewable energy technologies include solar panels, wind turbines, photovoltaic installations, heat exchange systems and micro-scale hydro-generation. Capital costs of the majority of renewable technologies are comparatively high at present, but as their use become more widespread economies of scale will drive costs down. Some governments such as the UK have grants under the Low Carbon Buildings programme to offset the initial additional cost and to provide incentives to the market.

Pollution from the construction process can take many forms other than the pollution into the atmosphere of greenhouse gasses; fuel spillages, fly-tipping and mud/silt from sites or lorry wheels are the most common. In addition, many construction materials can pose a pollution risk in their manufacture or in use. UK contractors can be prosecuted by the Environment Agency for breaches of the Environment Act (1995) and, if guilty, fined.

Life Cycle or Whole Life Costing is a technique that integrates the capital expenditure committed to a project with the operational costs involved in operating and maintaining the building. Clients procuring buildings will be disaffected if their development requires expensive maintenance soon after completion. Conversely, well thought out designs that incur little maintenance costs are to be commended.

Although using resources to their maximum efficiency is a good mechanism for achieving increased sustainability, it may not always deliver the best product for end-users. It is important to look to the end-users and make decisions based on utility, not only efficiency. The design and construction phases of development schemes are important here.

Waste management during construction can increase profitability and lower construction costs as wasted materials are paid when purchased and disposed of. In the UK around 13 million tonnes of the construction and demolition waste produced every year is made up of materials delivered to sites and never used (Environmental Agency, 2006). Sustainable property developers need to consider waste to save money and reduce environmental impact. Developers pay to dispose of construction waste, and landfill taxes and costs are increasing. A waste management strategy can help to design out waste, minimise waste creation on site and ensure any resulting waste is dealt with appropriately and this results in a tidier, safer site. In the UK some contractors use Site Waste Management Plans with reported savings of around 3 per cent of build costs.

Where developments involve demolition, a pre-demolition audit is necessary for effective waste management. Contractors and/or developers need to identify the type and amount of waste generated on site. In the UK a geographical information system (GIS) on the internet is used to reduce transport of waste by locating the nearest most suitable waste management site. Firms can locate nearby recycling sites, reclamation companies, composting facilities, manufacturer take-back schemes, transfer stations, landfill sites and incinerators. This is an emerging area and it is vital for contractors to measure and benchmark construction, refurbishment and demolition waste. The construction industry needs to establish minimum reporting requirements for construction, refurbishment and demolition waste and to generate appropriate performance indicators and benchmarking figures. With the imminent legislation for Site Waste Management Plans (for projects over a certain value), rising landfill costs and an increased attention to sustainability, it is necessary to start measuring and monitoring waste generated. Such practices will become the accepted norm; however, currently developers wishing to procure sustainable construction in their projects will need to set out requirements in respect of waste management explicitly.

Another way in which developers can increase the sustainability of a development project is to require the designers to specify the reuse of materials if the development involves the partial or complete demolition of a building. Also, the designer can specify recycled materials such as recycled concrete for hardcore or recycled timber. Reuse is better than recycling as no further embodied energy is used to transform the materials from one form to another. However, the use of recycled materials is preferable to the consumption of raw resources and materials. Environmental assessment methods such as BREEAM give credits to projects that incorporate recycled materials.

In respect of materials the following should be considered during property development:

- Environmental impact of materials. All materials have varying levels of environmental impact, for example excessive logging of timber can lead to deforestation and loss of the carbon sink.
- Responsible sourcing of materials, for example timber that comes from a sustainable source.
- Provision of recycling facilities for materials used within buildings during their life cycle; includes internal storage and provision for external collection.

Embodied energy is related to sustainability and is defined as the total energy required to produce the material or product, and comprises the total energy involved in extraction or mining, transportation and manufacturing of a material or product. Some materials such as concrete and steel have a high embodied energy whereas stone and timber are relatively low. If a developer wants a project with low embodied energy they should avoid using large amounts of high embodied energy materials.

Another consideration with regards to materials is the amount of VOCs content. VOCs are emitted as gases from certain solids or liquids. They include a variety of chemicals, some of which may have short and long-term adverse health effects. VOCs are emitted by a wide array of products including: paints and lacquers, paint strippers, cleaning supplies, pesticides, building materials and furnishings, office equipment such as copiers and printers, correction fluids and carbonless copy paper, and graphics and craft materials including glues and adhesives. Best practice avoids use where possible and limits exposure. Some individuals have a very low tolerance and health problems include eye, nose and throat irritation; headaches; loss of coordination; nausea; and damage to liver, kidney and central nervous system. Some organics are suspected or known to cause cancer in humans. Key signs or symptoms associated with exposure to VOCs include conjunctival irritation, nose and throat discomfort, headache, allergic skin reaction, nausea, fatigue or dizziness.

Water

Water shortages are becoming increasingly common in different locations and water needs to be used. In the UK the south-eastern corner of the country, the most densely populated area, is getting less rainfall and pressures are building on the water supply. This reduction in supply has coincided with increases in consumption. Developers can influence users to reduce consumption and running costs by including such measures as water efficient appliances (low flush toilets), water metering, leak detection systems and water butts. Environmental assessment schemes quantify what is considered reasonable consumption levels for the different property types.

Health and well-being

Sustainable buildings are promoted on the basis of benefits to health and consideration is given to optimising the health of occupants in the design of a project. These reasons are frequently cited as a good rationale for sustainability in buildings. In this respect developers should consider such features as:

- maximising natural daylighting;
- sound insulation to reduce the transmission of airborne and impact sound;
- in residential property, provision of adequate private space for occupants;
- not using materials containing VOCs;
- decreasing the amount of air conditioning and/or use of recycled air in large commercial buildings to reduce the likelihood of sick building syndrome.

Market research and sustainability

There has been a significant change in attitudes towards sustainability since the early 2000s. Surveys show that many people believe the environment to be very important. Marketing firms conduct primary consumer research to investigate the percentage of the population for whom environmental, social and healthy lifestyle values play an important role in purchasing decisions. Such research concludes that there is a shift in consumer attitudes towards sustainability and it is these general trends and shifts that filter down to property development. Property developers need to keep abreast of social and cultural trends, especially relating to sustainability, to ensure that their properties meet market requirements, reduce environmental impacts and are future-proofed to some extent. Surveys show around 23 per cent of the population has a profound sense of sustainability, 38 per cent has some sense of sustainability, 27 per cent possess a little sense of sustainability and the remaining 12 per cent have no sense of sustainability. Surveys also show that environmental benefits are now expected per se, for example people expect energy-efficiency in domestic appliances. Also, increasing attention to CSR has driven a move towards environmentally friendly products and services. Clearly these sentiments have filtered into property where the environmental impact is known to be considerable. Developers need to be aware of these changing attitudes and perceptions to sustainability in order to ensure their products meet market expectations and do not become victims of a new type of building obsolescence: environmental obsolescence.

Marketing sustainable developments

As stated in Chapter 7, developers need to identify their target market, work out the optimum ways of reaching that market and include the features that are attractive to that section of the market. It is not uncommon now to have two-tier marketing for residential projects where one set of brochures and campaign adverts is targeted at a more mature purchaser and another set that are focused on a younger, singles household group. Different sustainability features might appeal more to different groups, for example older people might prefer proximity to public transport or health care facilities whereas younger households might prefer access to recreation facilities like gyms and swimming pools. Generally speaking, marketing campaigns focused on sustainable schemes will seek to highlight the social and economic and environmental benefits of buying or renting the particular property. Furthermore, marketing campaigns might seek to infer that sustainable buildings will hold their value more as time goes by.

Promoting and selling sustainability

Increasingly, property developments are being promoted to the market and sold on their sustainability credentials. For example, in the residential sector, the volume house builders are selling developments as being sustainable because they are close to the city centre, energy and water efficient and so forth. Equally, with the commercial sectors such as offices, developments are promoted on the basis of their BREEAM assessment ratings. One of the stated aims of the BREEAM was to promote sustainable buildings in the marketplace. Given that many government departments have committed to minimum environmental standards developers are keen to promote their developments as complying with these standards.

On one level it can be argued that selling sustainability is an oxymoron, given that sustainability is about conserving resources and reducing environmental impact and that selling is about increasing consumption. It has been argued previously that selling sustainability features like energy efficiency was difficult because consumers were unable to physically see the features being promoted, whereas kitchens and bathrooms were very tangible to people. This appears to be diminishing as levels of awareness are increased about the benefits of low running costs. Such arguments are easier to make when fuel prices are volatile or increasing. So how can the two conflicting concepts be resolved? Some degree of development is inevitable, to cater for the changing needs of the community and changes in demographics and, vitally, to maintain economic activity. Therefore, it becomes appropriate and necessary to promote developments on the basis of their sustainability. It can also be argued that the promotion of sustainable buildings is highly desirable to further raise the awareness of sustainability in buildings to a wider audience.

As with any marketing and promotion campaign it is necessary to engage specialists to identify target groups and markets for the product. Advertising campaigns in a range of media can then be developed to target the groups effectively and create demand for the development.

The key property development stages are summarised in Table 9.1 with the main potential sustainability issues also noted.

9.9 Current initiatives and responses by government

Sustainability is a broad area covering many aspects of life, some affect property development directly and others indirectly. This section of the chapter looks at the UK government's response to sustainability in terms of policies and initiatives; where they directly affect property development some discussion is provided. The UK government launched its strategy for sustainable development in March 2005. The four priority areas identified for immediate action across the UK were:

1. sustainable consumption and production;
2. climate change and energy;
3. natural resource protection and environmental enhancement;
4. sustainable communities.

Table 9.1   Property development stages and the key potential sustainability issues

Property development stage	Potential sustainability issue
Land for development	• Loss of habitat • Loss of bio-diversity • Contamination (naturally occurring or due to previous use)
Development finance	• Consider eco-financing
Planning	• Transport • Ecology and site issues • Zoning and land use issues
Design and construction	• Selection of contractors – including their CSR performance • Reducing carbon dioxide emissions • Minimising pollution • Use of life-cycle costing or whole life costing techniques • Using resources efficiently • Waste management on site • Re-use of materials and recycling materials • Specification and selection of materials – health and embodied energy issues • Water • Health and well being for users • Environmental assessment ratings
Market research	• Awareness of changing social and cultural perceptions towards sustainability
Promotion and selling	• Awareness of changing social and cultural perceptions towards sustainability

The principles and approaches are covered in the UK Strategic Framework and in the government publication, 'Securing the Future – the UK Government's sustainable development strategy'.

Sustainable consumption and production is about doing more with less, it is about changing the way things are designed, used and disposed of in a move towards 'one planet living'. By this, the UK government is referring to the concept of the ecological footprint. The ecological footprint measures the impact of the way humans live in terms of the number of planets needed to sustain that style of living or consumption. Currently, the ecological footprint in developed nations is between three and five planets per person.

This initiative is looking at means of production, improving resource efficiency and encouraging sustainable consumption.

The UK government accepts that climate change is occurring; that it is caused primarily by human activity and that action is required to mitigate the effects of global warming if social, economic and environmental disaster is to be avoided. The UK has signed the Kyoto Protocol and has committed to reducing the UK's greenhouse gas emissions by 12.5 per cent below base year levels over the period 2008–12 and to reduce national carbon dioxide emissions by 20 per cent below 1990 levels by 2010. The UK Climate Change Programme (CCP) was published in 2000 and it sets out the policies to reach the goal in 2010. The policies include:

- promoting energy efficiency in homes and businesses;
- introducing business to the benefits of emissions trading;
- increasing the share of electricity generated by renewable sources;
- encouraging the take-up of less polluting vehicles;
- encouraging reductions of carbon dioxide emissions through the work of the Energy Savings Trust (EST).

In addition, an Energy White Paper launched in 2003, an Energy Efficiency Action Plan in 2004 and the EU emissions trading scheme have been established by the government to reduce energy consumption.

Under the terms natural resource protection and environmental enhancement, the government considers the natural resources vital to humankind's existence: namely water, air, soils and biological resources. They acknowledge that without well-functioning ecosystems the economy and key industrial sectors would face problems, though noting that these resources have intrinsic value regardless of functional value. The government collates data on: natural resources, air quality, biodiversity, fisheries, landscape protection, land-use planning –agricultural, land-use planning – general, marine environment, soil, water quality and forestry. A number of surveys and frameworks have been established covering natural resources. The UK government works independently and also in conjunction with international groups such as the United Nations Environment Programme (UNEP). Typically, much environmental legislation derives from EU directives, for example it covers air quality, noise, environmental assessments and protection, water and waste. Currently the EU's sixth Environment Action Programme (EAP) is running until 2012 and provides the key framework for policy development and objective setting. The UK government's sustainable development strategy falls within this framework. Indirectly, all natural resources affect property development; however, of particular importance to property development, are the policies and initiatives regarding land-use planning, both general and agricultural.

A key issue for the UK government is to modernise the UK planning system in a way that manages development and accounts for sustainable use of natural resources, for example by designing in measures to tackle issues such as flood risk, waste minimisation, recycling, water resource and energy efficiency. The planning system is vital to achieving sustainable development and the government planning policy statement 'Delivering Sustainable Development' (PSS 1) sets out the vision and policies to deliver sustainable development. The planning system needs to take into account the views of those affected by proposed developments. New planning guidance aims to develop a more integrated approach to management of resources. PPS 1 sets out the framework for reflecting the duty, in the Planning and Compulsory Purchase Act 2004, for regional and local plans to be prepared with a view to contributing to sustainable development. Other planning policies complement PPS 1 as follows:

- Planning policies for housing: ensure that brownfield land is developed first for new housing, and that new housing is built at higher densities than previously, reducing the need for use of greenfield sites.
- Policies: ensure new developments occur near town centres so that they are accessible by walking, cycling and public transport, reducing reliance on private motor vehicles.
- Policies: ensure finest rural landscapes are retained for public enjoyment.

Sustainable development occurs at all stages of the planning process. Regional plans, called Regional Spatial Strategies (RSS), are devised by Regional Assemblies and incorporate the Regional Transport Strategy (RTS), which together provide a spatial framework to inform the preparation of Local Development Documents (LDDs). The RSS should articulate a vision for the region for a 15–20-year time frame and illustrate how this vision contributes to delivering sustainable development objectives. RSSs and LDDs should be prepared in partnership with a wide range of stakeholders. The spatial policies within RSSs should cover:

- scale and distribution of new housing;
- priorities for the environment, such as countryside and biodiversity protection;
- transport, infrastructure, economic development, agriculture, minerals extraction and waste treatment and disposal.

As part of reviewing the RSS and LDDs, planning authorities must undertake a Sustainability Appraisal (SA) and guidelines have been provided. The SAs have to comply fully with the EU directive on Strategic Environment Assessment (see below).

The fourth key UK government priority area is that of sustainable communities, from global to local. The aim is to create sustainable communities in places where people want to live and work both now and in the future. The aim is to eradicate poor environmental quality and its impacts on health, high crime rates and unemployment levels. The issues covered here include many of the social aspects of sustainability, for example: sustainable communities, active communities, social exclusion, neighbourhood renewal, health inequalities, environmental justice, local transport, international development and millennium development goals. In all aspects, a range of policies and initiatives have been developed with the objective of reaching identified targets towards sustainable development.

Sustainability Appraisals

'Sustainability Appraisal' is a tool that provides for the identification and evaluation of the economic, social and environmental impacts of a proposal. In the UK SAs evolved in the 1990s from the environmental appraisal of development plans by local planning authorities and, while its application is now expanding, in the past it has principally been applied in the fields of local and regional planning. EU directive (2001/42/EC) requires Strategic Environmental Assessments (SEAs) to be completed on certain plans and programmes to assess their effects on the environment. This does not replace the requirement for certain plans to also undergo an SA. Under UK legislation (the Planning and Compulsory Purchase Act 2004), an SA must be prepared for regional and local development plans. It is similar to a SEA, but includes assessment of social and economic inputs, in addition to environmental inputs. An SA is used by planning authorities to assess whether proposed plans and policies meet sustainable development objectives.

While the requirements to carry out an SA and SEA are distinct, it is expected that in the future both can be satisfied in one single appraisal process. SAs should:

- take a long-term view of the expected social, economic and environmental effects of a proposed plan;
- check that sustainability objectives are turned into sustainable planning policies;
- reflect global, national, regional and local concerns;
- form an integral part of all stages of plan preparation.

Sustainability and Environmental Appraisals are processes that allow some or all sustainable development objectives to be integrated into policies, programmes, projects, activities and decision making at an early stage. They are designed to help identify potential environmental, social and economic effects and issues as early as possible, allowing alternative solutions or mitigation measures to be explored. Alternatively, positive effects and opportunities for performance enhancement can also be identified and promoted. These methods allow sustainability and environmental issues to be considered in a systematic, transparent and auditable way.

Department for Communities and Local Government

This is the key UK government department covering property development issues. The UK government sees its key contributions to sustainable development as:

- creating sustainable communities that embody the principles of sustainable development at the local level;
- providing homes for all, while protecting and enhancing the environment;
- working to give communities more power and say in the decisions that affect them; and working to improve governance at all levels so that we can work at the right level to get things done;
- creating cleaner, safer, greener agenda: to make public spaces cleaner safer and greener and to improve the quality of the built environment in deprived areas and across the country by 2008;
- promoting sustainable, high-quality design and construction, to reduce waste and improve resource efficiency, and promoting more sustainable buildings;
- putting sustainable development at the heart of the planning system, as set out in PPS1 'Delivering Sustainable Development'.

Their strategies for delivering their policies are contained in the 'Department for Communities and Local Government – Sustainable Development Action Plan', which is available online at http://www.communities.gov.uk/publications/corporate/sustainabledevelopment2007-08.

Carbon neutrality

This is an emerging area where businesses aim to have zero net carbon dioxide emissions and use a range of options to achieve the target such as reducing carbon emissions, using renewable energy and offsetting the remaining balance of emissions. The government aims for its office estate to be carbon neutral by 2012. Emissions that cannot be eliminated can be offset either through carbon reduction projects overseas or by paying into a fund to reinvest in energy efficiency on the government estate. Carbon offsetting should not be seen as the ultimate solution for carbon emissions, but it does provide a way to mitigate the remaining emissions. It is likely other businesses will seek to become carbon neutral and this is likely to affect the design and operation of property developments. Furthermore, the concept of designing carbon neutral buildings is becoming more widely discussed.

Joint UK–Sweden Initiative on sustainable construction

This initiative shares best practice in sustainable construction from both the UK and Sweden and the goal is to encourage sustainable construction, management and renovation. The building and property sector is responsible for a very high proportion of environmental impacts. Environmental improvements within this sector are vital in achieving a sustainable society. The website helps to foster new joint UK–Swedish projects using case studies, or by putting forward building or refurbishment projects that need or would benefit from international partners.

Building regulations

The building regulations are a key mechanism for government to enhance sustainability in the built environment. In April 2006, revised Part L, Conservation of Fuel and Power, of the building regulations came into effect; the changes will improve standards by 40 per cent. The improved standards mean that developers will need to make greater use of insulation, efficient boilers, and to consider zero or low carbon systems such as solar and wind turbines to achieve compliance. The adoption of air pressure testing also makes the energy certification process required by the EU Energy Performance of Buildings Directive (see below) easier. Apart from energy, building regulations can also cover other sustainability issues such as water economy, waste and recycling, for example. It is anticipated that in the near future these measures will be either enhanced or introduced into the UK system.

EU Energy Performance of Buildings Directive

The EU Energy Performance of Buildings Directive (EPBD) was published in January 2003. The overall objective of the Directive to:

> promote the improvement of energy performance of buildings within the Community taking into account outdoor climatic and local conditions, as well as indoor climate requirements and cost-effectiveness.

This legislation represents a major change in that building owners are required to identify, monitor and display the energy performance of their buildings. Property developers will have to ensure that their buildings meet standards if they are to appeal to the market. Each member state was required to transpose the directive into law by 2006, with a further 3 years allowed for full implementation of specific articles. The Energy Performance of Buildings (Certificates and Inspections) (England and Wales) Regulations 2007 amends part of the Building Regulations 2000 and sets out duties to produce and display energy performance certificates; duties regarding advisory reports; inspection requirements for air conditioning; energy assessors and accreditation schemes; registration of certificates and reports; and enforcement and penalties for non-compliance. This legislation forms part of the UK climate change programme to promote energy efficiency in property. Article 7 requires the production of energy performance certificates (EPCs) and advisory reports when buildings are constructed, sold or leased and to display certificates in large buildings occupied by public authorities and some public institutions. The legislation states that EPCs must not be more than 10 years old.

Property development-related environmental assessment tools

There are different assessment tools available and developers need to identify which tools suit their needs best. Some tools focus on single-issue aspects such as energy efficiency, and the National Home Energy Rating (NHER) is an example. Other tools cover a broader range of issues and the BREEAM is an example of this type. The tools provide a rating that benchmarks the project's sustainability, environmental impact or energy efficiency, depending on the type of tool used. The ratings can then be used to promote and market the scheme to potential buyers and users. Some of the environmental assessment tools currently used in the UK and Europe are identified below.

Since 1990, BREEAM has provided a broad-ranging assessment of the environmental impact of a building under the headings of global, local and internal environments. BREEAM can apply to the design stage (i.e. new build and refurbishment) and to the operation and management of a property. The scheme covers a broad range of building types such as prisons, hospitals, homes, multi-residential homes, offices, industrial, retail and educational buildings. Ecohomes provides a rating for new, converted or renovated homes, and covers single dwellings and apartment buildings (BRE, 2007).

Office Scorer is a tool that evaluates major refurbishment with total redevelopment, and redevelopment within an existing facade. It enables users to contrast and assess the environmental and economic impact of different building designs for offices and to identify sources of further guidance.

The BRE Green Guide to Specification provides guidance for specifiers, designers and clients on the environmental impacts of over 250 elemental specifications for walls and floors etc. Environmental ratings are based on Life Cycle Assessment using the Environmental Profiles methodology developed by BRE with support from construction and manufacturing trade associations. Manufacturers can obtain certified Environment Profiles, which allow products to be compared to generic profiles (BRE, 2007).

Envest 2 is a software tool that makes straightforward the process of designing buildings with a low environmental impact and establishes whole life costs. It permits environmental and financial trade-offs to be modelled in the design stage, allowing a developer or client to optimise best value.

There are sustainability checklists for the design phase of a project that include the BRE sustainability checklist. The checklist allows developers, planning authorities and advisors to specify and assess the sustainability measures of developments. It comprises a series of steps that can be followed to incorporate sustainability into developments, and reflects latest guidance on sustainability. Where possible, it uses existing systems and standards to define performance, such as BREEAM, EcoHomes (the homes version) and 'Secured by Design' (Rao *et al.*, 2004). The South East England Development Agency (SEEDA) sustainability checklist is an assessment tool that enables the sustainability aspects of a property development to be addressed, and for reviewing organisations such as local authorities, SEEDA and Government Office of the South East (GO-SE) to understand the level of performance that might be achieved. It is used by developers to show the sustainability credentials of their project.

Environmental Impact Assessment (EIA) and Strategic Environmental Assessment (SEA) are tools that comprise a group of processes to evaluate sustainability. EIA concentrates on the construction and operation of large development projects, such as airports. It is focused on environmental impacts, though the definition of environment embraces social aspects affected by the environment, for example human health. Globally, EIA is well developed, and is necessary for large projects in the UK. SEA, on the other hand, is more centred on qualitative predictions and on techniques that deal with uncertainty, such as the types and locations of project that might arise, and the likelihood and types of new technologies emerging.

There are some industry initiatives such as SPeAR, an environmental indicator assessment method developed by Arup Environmental to help companies evaluate and improve the sustainability of their products, projects or performance. SPeAR examines natural resources and the environmental, economic and social sustainability issues and produces a diagram showing the negative and positive effects of the project. The indicators are adapted for individual projects but may feature some core indicators. These indicators can be used at any point during the project to measure and monitor performance. The appraisals identify areas where a project/design or development is performing well or can be improved in terms of sustainability. For instance, positive effects might be low energy consumption using renewable energy, whereas poor performance could include the negative impact on local employment opportunities. It is not known how widespread the use of this tool is and there is an issue with adopting assessment methods that are less well known and accepted in the marketplace, which developers need to take into account.

Finally, there is nothing to stop a property developer getting a bespoke assessment for a scheme. Many ecologists would argue that these projects are likely to be more sustainable than those in the 'off the shelf' suite of products. This is because designers and developers do not become fixed on targets established by others as being the 'standard'. On the other hand, it can make it less easy to promote and convince the market that the bespoke building is as or more sustainable because the recognised benchmarks have not been adopted. Bespoke assessments may be more useful to those developers or owners wishing to retain a long-term interest in the building or scheme.

Discussion points

- What are the main environmental assessment tools available to property developers?
- Why would property developers use these tools?

Reflective summary

We are now in an environment where developers ignore sustainability at their peril. This chapter started with an overview of sustainability and how it relates to CSR. The question of why property is an important sector in relation to sustainability and CSR was addressed and the key areas of impact were identified. Finally, the chapter addressed how these impacts are being addressed by government and by developers, investors and occupiers.

It is impossible to ignore sustainability as some issues are embedded within legislation such as planning and building regulations, for example energy efficiency and water economy. However, developers should be aware that these regulations represent the minimum allowable standards and not best practice. Developers should adopt best practice wherever possible. The standards and quality of the built environment should be improved for both inter and intra-generational benefits. Sustainability can be embedded throughout the development process, from inception to site selection and acquisition to the financing of the scheme. The design and procurement phase is another key area where decisions will have a substantial impact on the sustainability of a project. Environmental assessment tools are growing in their scope and range and a brief overview of the key tools used in the UK was provided. These tools enable benchmarks to be set, which the market recognises and acknowledges. The importance of market research is highlighted with the need to consider the different types of sustainability features that would appeal, especially to the target groups that a particular project would be aimed at. In all areas of property development and at all stages in the process the trends are for more sustainability. Over time the tools adopted in industry are modified and improved thereby allowing developers to deliver buildings to the market that are more sustainable than previously.

Case study: Red Kite House, Howbery Park, Wallingford, Oxfordshire, UK

Background

Red Kite House is a new office for the Environment Agency (EA) in Wallingford, England. It is an example of a sustainable property development and demonstrates leadership in the office design. Working closely with the developer, the EA agreed to take a 15-year lease of the building on the proviso that they could influence the building design. The developer and the occupier agreed to construct an office that would meet the operational needs of the occupier and demonstrate best practice in sustainable office development. The EA required that the building achieve a BREEAM rating of 'excellent'.

Red Kite House is a three-storey, naturally ventilated office building with an internal floor of approximately 3,000 m². Moss Construction, a division of Kier Regional, constructed the building for approximately £4,500,000 in 2004 on a design and build contract. The design and build contract enabled the developer, contractor and occupier to work together closely to achieve a successful outcome, especially in respect of sustainability issues.

Plate 9.1   Red Kite House, Wallingford, UK. Reproduced with permission of King Sturge (Source: http://www.rics.org)

Table 9.2   Summary of sustainability savings and benefits

Item	Consumption saved per annum	Tonnes of carbon dioxide saved per annum
PV cells that generate solar electric power	23,000 kwH	12
Solar heating panels for hot water	3,100 kwH	1.6
Rainwater harvesting system	240,000 litres of water	n/a
Natural ventilation as opposed to air conditioning	7,500 kwH	4
Total	33,600 kwH 240,000 litres of water	17.6

Source: adapted from King Sturge, 2007

Table 9.3   Red Kite House – approximate costs

Item	Developer	Tenant	Grant	Total after grant
Base building, including permeable car park membrane	4,100,000	0	0	4,100,000
Canopy that generates electricity from PV cells	0	295,000	165,000	130,000
Solar heating panels for hot water	0	23,000	9,000	14,000
Rainwater harvesting system	0	35,000	0	35,000
Roof mounted fans to support natural ventilation	0	10,000	0	10,000
Automated and motorised high-level windows	30,000	86,000	0	116,000
Internal fittings and furnishing	75,000	1,400,000	0	1,475,000
Total cost	4,205,000	1,849,000	174,000	5,880,000

Source: King Sturge, 2007

Sustainability features

Some of the sustainability features include in the build are identified below.

Energy

- The photovoltaic (PV) cells on the south-facing canopy generate electrical power and reduce carbon dioxide emissions by around 12 tonnes per annum.
- Hot water is produced by solar thermal panels reducing carbon dioxide emissions by around 1.6 tonnes per annum.
- The building in operation is expected to have carbon dioxide emissions about 26 per cent lower than that defined by government as 'good practice'.

Natural ventilation and cooling

- The building is curved in shape, naturally ventilated and orientated to capture the wind, maximising air flow and cooling.
- Natural cross ventilation of the building is provided by manually opening windows.
- High-level exposed concrete ceilings on each floor provide a heat sink during the day. They are cooled by air entering through a hundred motorised windows on each floor.
- Roof-mounted turbines draw air in through the top-floor windows, which is the most vulnerable floor to overheating during the summer period.
- Neutral solar control glazing minimises solar gain in summer while maximising natural daylighting.
- A south-facing canopy provides external shading, especially to the top floor.

Water and drainage

- A rainwater harvesting system collects and reuses water, meeting around 40 per cent of the buildings yearly requirement for water. Overspill from the tank is directed to a reed bed.
- Pervious blocks are used in the car park to allow rainwater to soak into the ground rather than run off into watercourses. Oil and other pollutants are trapped by a geotextile membrane below the blocks.

The information for this case study was obtained from European Property Sustainability Matters published by King Sturge, London, and is reproduced with their permission.

International practice

10.1 Introduction

This chapter explains how property development has changed recently, partly as a result of the globalisation of business and the widespread adoption of information technology including the internet. The concept of globalisation is described and then explained within the context of property. The opportunities for international property development are discussed, along with the drivers for international property development. This chapter provides an overview of the opportunities, barriers and risks associated with international property development. An outline of an international strategy is provided prior to moving onto examples of international property development.

10.2 Background

Globally, there has been a move to increase the number of people living in city centres. Partly as a result of urban planning policies and high land values, many city centres in the 1990s became devoid of residential components causing many inner cities to become lifeless areas outside working hours. Urban regenerative city centre living is an example of some of the initiatives promoted in numerous cities in developed countries around the world over the last decade or so. These policies and initiatives have lead to growth in the inner city residential apartment block in global cities ranging from San Francisco to Toronto and Sydney to Manchester. Furthermore, in some cities during economic downturns or periods of high vacancy rates in commercial building stock there have been conversions of high-rise property from office use to residential use. This is a complex change of use and until the mid-1990s was a relatively uncommon one. Interestingly, however, it became an option that was picked up and emulated in Canada, the US and the UK and demonstrates the internationalisation of property, development practice and property markets.

Another feature of changing practice internationally in the property industry is the relatively rapid growth in purchasers of investment property 'off the plan' from overseas investors located in another market. These purchasers are individuals and not large investment funds. For example, many ocean-front locations in Australia are often promoted internationally to overseas investors living in Asia. At times, this can result in a two-tiered market consisting of local buyers and overseas buyers, and an international property development should be marketed accordingly to maximise returns. In the EU, for example, many European property transactions have occurred where 'cashed up' baby boomers have been looking to purchase holiday homes or second homes as retirement properties that are sound investments. In addition, the internet has made it easier for overseas investors to research property in other countries and then purchase it, often sight unseen. At the same time there has been growth in the number of property agents willing to facilitate these international property transactions.

In the era of globalisation an economic event in one country can have an adverse and almost immediate effect on other global markets, especially when referring to exchange rates, stock markets and interest rates. Since the 1980s, property cycles have been increasingly caused by various conditions in the international economy, but also have been influenced by national macroeconomic conditions (Dehesh and Pugh, 2000). More than ever before business is now conducted in an increasingly smaller world with many companies operating on different continents, where the location of the head office is often difficult to quickly identify. Technology such as the internet and reliable (and expensive) communication channels have assisted greatly here. Thus, these companies, especially construction companies and property developers, have diversified their primary geographic location of operation in order to benefit from enhanced profit in accordance with conventional economic theory.

In countries such as India and Bangalore the pursuit of capitalism and free market trade has led to the massive expansion of outsourced labour, such as an increase in the number of call centres. In other words, international companies such as telecoms are able to provide telephone help lines at a much lower cost due to cheaper labour and infrastructure costs. This in turn this has caused an expansion in demand for property development in these countries, especially for high-tech office space, which previously was not the standard level of accommodation in these countries.

Previously, local real estate markets were always viewed as influenced primarily by local economic activities, although with the increasing acceptance of internationalism this has changed over time. However, property developers have not benefited from entering global markets as much as other organisation types, even though investment in and development of real estate has traditionally been linked to wealth creation and achieving investment goals. In the past there have not been any property development companies in the largest 100 transnational companies ranked by foreign assets in the world; this list is dominated by oil, electronic and automobile companies (Hailia, 2000).

In the context of property development it is no longer the local company that would most likely always be the successful bidder. After consideration is given to the scope of the property development, such as a football stadium or a detached house, the developer is not automatically located in the immediate or local vicinity. There is a need to develop an understanding of property development processes that combine a sensitivity to the economic and social framing of development strategies with 'fine-grain' treatment of the local contingent responses of property actors (Guy and Henneberry, 2000).

Following the integration of technology and the internet throughout the western world, many of the geographical boundaries that previously existed have now been substantially broken down. For example, it is practically impossible to determine if the person you are talking to is located in India, UK or the USA. The head office of a property development project may be located across town, across the country or across the world and still be able to compete directly with a local provider. Such increased levels of international practice have caused higher levels of competition, where a truly international property developer must be fluent with the local customs, government regulations and state of the market in many different areas. Similarly, consultants who service property developers, such as services engineers, architects, agents and quantity surveyors for example, have also internationalised. Sometimes this has taken the form of mergers and acquisitions of local practices, whilst at other times partnerships have been formed. This expansion and internationalisation has meant that consultants who work on projects in one country may find themselves working together in project teams in other countries.

It is necessary to discuss the concept of globalisation because it has had a substantial impact on international property development, and it is an impact that looks set to increase. Globalisation refers to the increasing worldwide connections, integration and interdependence in political, economic, environmental, social and technological areas of interest. The *Encyclopaedia Britannica* (2007) states that globalisation is the 'process by which ... everyday life ... is becoming standardised around the world'.

Globalisation is an undeniable, established and irreversible trend as a result of the lowering or removing of political and trade barriers, the advent of fast, safe and cheap air travel, rapid development in new technologies, and a very high level of productivity from emerging economies such as India and China. Not only has the volume of world trade increased over the past 25 years but it has increased at an accelerating rate. For example, according to the World Trade Organisation (2007) total world merchandise trade for exports in 1996 totalled US$ 5,402,000 (millions), a decade later in 2006 this figure had grown to US$ 12,062,000 (millions). Historically, when large amounts of capital flowed across national borders it was usually deposited into the equity or government bond markets, now, however, there is a global real estate market in which to invest.

There is a debate about whether globalisation results in a homogenised society and culture with converging patterns of consumption or whether society can retain individual cultural differences in the long term. It has meant that identical products and services are found in an increasing range of countries and this includes property developments. There is certainly evidence that architecturally there is increased similarity with property in our world cities, for example office buildings in Toronto look much like office properties in Sydney or Kuala Lumpur. Such practices are further encouraged by global companies insisting that their branding is adopted globally. For example, the food retailer McDonald's uses similar design and fit-outs for its properties throughout the world. Another example is international hotel chains, which use identical specifications in their buildings.

Economically, globalisation is about the convergence of products, wages, prices, interest rates and profits towards the norms in developed countries and the International Monetary Fund (IMF) has commented on the increasing economic interdependence through the growing amount of cross-border transactions, international capital flows and more rapid and widespread diffusion of technology. However, globalisation has a wide range and number of impacts, which are summarised as:

- industrial
- financial
- economic
- political
- informational
- cultural
- ecological
- social
- technical and legal

All of the above impacts can affect property developments. For example, the cultural impact of globalisation is that there is a raised consciousness and awareness among the peoples of the world who desire to consume foreign products and ideas. People nowadays want to participate in world culture and increased overseas travel has added to the phenomenon. For example, in Australia European kitchens are considered very sophisticated and upmarket in residential property developments and the marketing and promotion of the developments reflects these specifications. Equally, many developments market aspects of feng shui. This is the ancient Chinese practice of placement and arrangement of space to achieve harmony with the environment. Feng shui is a discipline with guidelines that are applied in architecture and property design; advocates claim feng shui affects health, wealth and personal relationships. Some property developers are promoting and marketing their projects stating that the principles of feng shui have been adopted in the design of the scheme.

The result of globalisation has been that many large corporations have become transnational firms where companies focus on staying competitive by outsourcing services or production to developing countries with very poor labour, environmental and economic standards. Such business practices allow companies to economise, leading to larger returns for investors and cheaper services and products for consumers. There is an argument that such practices encourage the governments in developing nations to retain poor labour and environmental legislation. Some transnational corporations lobby governments to gain entry to developing countries, though the advent of CSR is beginning to curb some of their less than exemplary overseas activities. These companies seek to avoid any negative publicity that results from associations with poor business and environmental practices in developing countries.

However, globalisation is complex and countries do seek to maintain their economic advantage. This means that many developed countries still have protectionist policies that prevent developing countries from exporting to developed markets. The result is that these trade barriers impede economic development in the Global South and not the Global North. Critics of the expanding global economy argue that the reduction of trade barriers will create harder competition for the previously protected companies in developing countries, while its advocates point to the new possibilities in global markets for emerging market companies such as Haier (China) and the Tata Group (India). In 2006 the global economy continued to expand so that most individual, corporate and government borrowers are making good on their obligations, which in turn has kept financial markets performing well and, thus, property developers are working to meet market demands for new and/or improved facilities.

10.3 Globalisation of property development

So what does globalisation mean for property developers? According to the Global Investment Strategy section of LaSalle Investment Management, capital flows in real estate around the world in 2005 were US$700 billion, 20 per cent higher than in 2004. US $126 billion – 18 per cent of the total – was in cross-border transactions with Asian countries, which was the top market for foreign investment. To establish some context, US$700 billion is equivalent to 700 buildings of 1 million square feet each at a unit price of US$1000 per square foot. This is the volume of real estate that changed hands across national borders in a single year.

On a more national scale this level of activity has meant that many UK-based property developers now operate in countries other than the UK. Some have established second or satellite offices in other countries whereas others operate solely from overseas countries and no longer have UK operations. There has been a substantial growth in Britons investing and purchasing property outside the UK and these developers are taking advantage of the opportunities presented. UK buyers overseas are generally happier dealing with UK nationals who are able to explain the process of property acquisition and development in an overseas setting. The expansion of this area of property development business has been assisted by the growth and economic integration of the European Union, which has lead to much increased pan-European activity in the last decade. Similarly, property developers from countries outside the UK will operate within the UK if the economic circumstances and business opportunities present themselves.

When it comes to the market for overseas property buyers and investors, take the case of Britons. Consider residential investment properties or retirement/second homes for individuals as an example. The UK Office of National Statistics (ONS) has the most reliable figures about UK ownership of properties outside the UK. However, even their figures are not perfect as they take account only of the number of households owning property overseas and do not take account of households owning more than one property overseas, and that is a sizeable number. The ONS produced figures on the state of the market in 2003/4 and these showed 231,000 Britons owning property overseas with a value of £23 billion. This figure had increased from £7 billion in 1994/5, an increase of 228 per cent in 9 years. The average price of property, estimated by the ONS and using statistics from the European Union, placed the average value of a British-owned overseas property at £99,567 in 2003. The 2006 report shows the average price as £98,166.67. Three years on and do these figures mean that prices have not gone up in that time? Rather than showing no appreciation in property prices in that time, it actually reflects the emergence of new markets with significantly lower property prices such as Bulgaria and other former Eastern Bloc countries. Furthermore, the increase in Britons working overseas selling and promoting these developments encourages buyers because there is a sense that they will trust people from their own country. If Britons buying and investing property abroad were dealing solely with local people they would come across language and cultural barriers, which would deter the more cautious buyers from completing the deal.

To cater for transnational businesses and cross-border developments there are three key ways in which property developers can obtain general economic and property market information. First, an effective way is to use international property consultants, many well known UK-based property companies now operate in countries outside the UK. They have local employees who know their market well and can advise developers on all aspects of property development in their locality. Furthermore, these consultants have access to networks of other allied consultants that can assist with the process.

Second, there are also companies that provide independent, accurate, comprehensive and up-to-date research on industries operating in their country. The data includes reports providing statistics, analysis and forecasts. These companies may prepare reports on the top nationally performing companies and provide reports with risk ratings on different industries. Market research studies investigate investor intentions and report that, for example, more than 60 per cent of US real estate executives, investors and other experts expect to invest in properties or land outside the country in 2007–8. The Bryan Cave Real Estate Executives Forecast Survey found that 61 per cent of real estate professionals plan to park cash outside the US, with the greatest interest being in Mexico and China. The survey, which polled 343 professionals, including public and private real estate company executives, investors, opportunity funds, commercial mortgage bankers, lenders and brokers, showed that 15 per cent consider Mexico as a key investment market, while another 15 per cent named China. Other countries of interest included the UK, Canada and Japan, where 12 per cent, 8 per cent and 8 per cent, respectively, of those surveyed plan to invest in land or property. Brazil, Mexico, Russia, China and India will produce five times as much real estate in the next 10 years as will be created in the US. Access to such data illustrates where opportunities will lie for international property developers. This type of information will enable developers to make more informed and, therefore, better business decisions.

Finally, a third way for developers to find out about market conditions and opportunities in countries outside the UK is to use affiliated professional bodies where available. For example, RICS is a global professional body representing land, property and construction. Members of RICS operate in many countries outside the UK and can offer a range of professional advice and services to property developers.

10.4 Opportunities

Many property developers who are examining the global market are seeking opportunities to identify markets that have potential for future growth or, alternatively, are currently under-developed. In many instances this will require the developer to be an early adopter in the marketplace, rather than waiting until the market reaches maturity with many property developers competing for the limited supply of prospective sites. Hence, a developer with foresight to enter a growing market has the benefit of rapidly establishing goodwill and strong links with the local market, as opposed to entering a competitive mature market from a standing start.

At the same time as property developers have expanded globally, there has been a parallel with increased global investment. Since ownership regulations and differential taxation are not large barriers and sources of segmentation in different property markets, many larger buildings in developed countries are owned by international investors. It should be noted that the mix of international investors change over time. The UK property sector has been the destination of substantial portfolio investment, with the main investors from Japan, the US, Sweden, Germany, The Netherlands, the Middle East and Asia, where the devaluation of the sterling in 1992 was associated with a sharp increase in investment by German funds (McAllister, 2000).

At times, other countries will offer opportunities for higher returns than the host country of the property developer, although this will vary according to property cycles and conventional supply and demand interaction. Such opportunities for overseas companies in the construction industry in Asia were identified in the 1990s. The main trends observed were:

- larger private sector participation in infrastructure projects;
- increasing vertical integration in the packaging of construction projects;
- increased foreign participation in domestic construction (Raftery *et al.*, 1998).

Since then such trends have continued in Asia and many overseas companies have benefited from the Asian development boom, especially in China. The gradual 'opening up' of previously 'closed' economies has been encouraging news to international property developers, although caution should be exercised before undertaking a large capital commitment without conducing adequate due diligence. For example, in many countries there are still difficulties in accessing reliable and timely detailed property data, such as the volume of sales and actual transfer prices. Some countries still have their supply of property largely controlled by the government and this should be carefully monitored. Operating in an environment that does not operate freely can increase the exposure to risk.

International property development affects each country in a different way. For example, a developing country's fledging firm may confront formidable competition from transnational corporations possessing far greater economies of scale, and global networks (Hill and Kim, 2000). However, the same argument is support by the belief that less developed countries have hidden reserves of labour, savings and entrepreneurship. These strengths and weaknesses should be identified by the progressive property developer, thus opportunities do exist in many countries with each in a varying stage of development itself. For example, partnering has been occurring for quite some time between British and Chinese firms in some Asian countries (including China), and this trend is likely to continue in the future (Pheng *et al.*, 2004).

In certain countries geographical clusters have formed. Cluster development has rapidly developed in North America, Europe and newly industrialised countries – clusters are located around the globe from Scotland to Bangalore and from Singapore to Israel, although many can be linked back to the original Californian model of Silicon Valley (Wolfe and Gertler, 2004). Cluster development has attracted the attention of different countries seeking to establish an identity, which in turn may provide an opportunity for an international developer that is able to bring specialised skills. Clearly each country, and in particular each region in a country, can promote a limited number of clusters. In this example a specialist property developer will have the competitive advantage of assisting to develop a cluster based on their previous experience in other countries.

A driver for establishing global links is the desire to be recognised as a truly international property company. There are many organisational models possible, ranging from a management structure with the majority of the workforce sourced from the local economy, to relocating the entire workforce from another country, with associated relocation and housing costs. Initially, the property developer will seek to reduce costs as well as exposure to risk, therefore, relocating the minimum number of workers required for the initial property development will often be the preferred option.

Discussion point

What are some of the drivers behind the expansion of property development into other countries?

10.5 Barriers and limitations

The transition to a truly international property market has been faced with challenges, especially when considering differences in currency, culture and varying levels of development in each country. As some real estate markets move through the transitional stage to a truly market-based structure, it is important that valuations (i.e. the business case for property development) align with the expectations and requirement of the international investment community (Mansfield and Royston, 2007).

Lack of knowledge about the inside workings of a property market can be one of the largest barriers to a successful property development in an overseas country. In contrast to the type of information that is freely available in general circulation in many developed countries, especially in terms of reliability, accessibility and cost, property-related information may simply not be as available as anticipated. Even if the information is available, there may be a premium attached for 'non-locals' and also questions raised about the level of reliability. When considering the high importance placed on the final sale price of the property development, some companies venturing into overseas countries have been disadvantaged by this lack of local knowledge.

Cultural barriers may exist, for example if a property company used to conducting business in a developed country seeks to develop property in a country with a transitional or emerging economy. Depending on the different cultures it may take time to establish a strong trust between the property developer and the local property stakeholders. It has been demonstrated that trust-based relationships create advantages in conducting business such as lowering cost, shortening duration and improving performance (Bromiley and Cummings, 1993). These are all attributes that a property developer in an international marketplace is keen to improve upon. For example, the Chinese culture comprises certain core values such as trust and *guanxi* (relationship) that influence business operations; in such an environment it may be essential for project participants to know what risks are inherent in relationships, and what tools for fostering trust and managing risks should be adopted (Jin and Ling, 2005). On the other hand, it has been argued that Chinese contractors are often accused of poor performance and low effectiveness in terms of quality and performance (Sha and Lin, 2001).

Language and cultural barriers can vary significantly between countries, although to varying degrees. For example, in one study it was shown that there were no signs that differences between Swedish and UK company cultures have an adverse impact on knowledge transfer (Bröchner *et al.*, 2004). The same study concluded that rich media, for example video-conferencing, and, preferably, face-to-face meetings are perceived to be good for the transfer of knowledge; leaner media such as email was also seen to be efficient. In other words, language barriers can be broken down using media and technology including email and teleconferences. Hence communication with a distant geographical location can be maintained on a regular basis if accompanied by adequate forward planning and organisation.

10.6 Risks in international property development

The increasing internationalisation of property markets has increased the level of demand for property, although arguably it has at times exposed property investors to more risk. One view is that the international construction industry can be characterised as highly volatile, subjecting contractors to financial and geopolitical risks (ENR, 2003).

For example, after the Paris office market fell by two thirds in value between 1990 and 1995, North American hedge funds were the first to enter the market and were skilled in investing against the business cycle (Nappi-Choulet, 2006). These funds then benefited from large capital gains from 1995–9, although this was closely followed by a 40–50 per cent increase in rent for prime office space. This example highlights the risks associated with understanding demand for property in an overseas market, especially with respect to planning an office development. Say, for example, if a developer examined the Paris office market prior to 1990, between 1990 and 1995, or after 1995 there would be three completely different development situations. After factoring in the external influence of the hedge funds in this example, it can be easily seen that a full appreciation of the international market is essential.

An international property developer may be exposed to different types of risk. Structural risk may come from within the property development industry itself, although growth risk may occur from the anticipated growth of the overall property market. Also, external risk can result from forces external to the development, but once again outside the control of the property developer. It is accepted that risk can not be fully removed from the project, although identifying the varying types and levels of risk will assist in understanding the threats to successful completion of the property development.

It is accepted that establishing trust between the property actors is absolutely critical when seeking to undertake property development. The importance of this relationship is emphasised when the actual property development site is located at a remote location, or at least not in the vicinity of the head office. It has been shown that as a relationship with an international partner progresses, more inherent risks are produced – in these instances trust-fostering tools are required (Jin and Ling, 2005). Furthermore, as these tools are applied in different stages of the property development, trust develops, which in turn counterbalances the risks. This, then, reduces the risks associated with achieving a successful and timely completion of the project.

Many regions promote a 'buy local' culture, which may indirectly create barriers for companies that are perceived as 'outsiders'. To overcome this barrier it is important to consider the culture and environment of the area surrounding the property development. Employing local workers and subcontractors, where possible, will partly address these concerns. For large projects a public relations expert may be employed to ensure interested stakeholders are kept up to date with important facets of the development, as well as emphasising how much the project will contribute to the local economy.

Another approach for addressing local responsiveness is to organise a strategic alliance with a local partner who is located in the region. The strategy presented in Figure 10.1 is designed to highlight alternative approaches to balancing global expansion with the local market. Pressures for global integration occur when a property developer is selling a standardised good or service with little ability to differentiate its products through features or quality. According to Griffin and Putstay (2007) the four strategies are as follows:

1. **Global strategy**   occurs when pressures for global integration are high but the need for local responsiveness is low, such as the expansion of Japanese consumer goods into global markets.
2. **Transnational strategy**   is when both global integration pressures and local responsiveness pressures are high, such as a producing a worldwide motor vehicle, albeit designed to meet local market specifications.

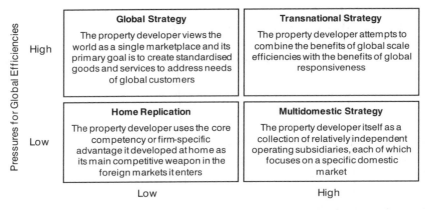

Figure 10.1 Strategic alternatives for combining global integration and local responsiveness (Source: adapted from Ghoshal and Nohria, 2003)

3. **Home replication**   is adopted when pressures for global integration and local responsiveness are low; an example being a retailer who sells the same commodities successfully to all global markets.
4. **Multidomestic strategy**   occurs when the response to local conditions is high but pressures for global integration is low, such as where global producers sell a product known worldwide at a premium to the local market using local market resources.

Currency risk is an important consideration when undertaking an international real estate venture (Sirmans and Worzala, 2003) or when involved in a property development across international boundaries. Although careful analysis can be given to the standard risks associated with a standard property development, operating in another country can dramatically increase exposure to new risk factors such as currency risk. Other indirect risks can also be adversely affected when working in a foreign country, such as transport risks to a distant location where there is a heavy reliance on air travel. For example, travel by air is relatively expensive and may be affected by irregular services – in contrast to a local property development readily accessible by road, air travel is subject to flight schedules and availability, with associated waiting periods of days. An urgent crisis may suffer adversely if the correct person can not be on site due to transport risk.

The relevant legislation and political climate are also major considerations that may hinder a prospective international property development. Again, these factors are completely outside the control of the developer and careful research should be undertaken prior to entering the marketplace. After completing extensive research, if the cost–benefit analysis indicates that the proposal is not viable due to the added risk involved, the project should not be pursued. Exceptions to this decision may occur, such as when the company is willing to accept a higher level of risk in order to make a strategic, high-profile entry into the market with a long-term perspective. The added risk should be factored into all aspects of the analysis (i.e. not only the overall level of profit and risk), including sourcing of local labour and local materials, timely completion of the project as well as demand for the finished property development, i.e. either rent or sale.

There is a view that because of the changes to the weather patterns and changes in sea levels induced by anthropogenic activities there will be a need for considerable international infrastructure and development in the forthcoming decades. Rising sea levels will create the need for improved and/ or new flood defences or even relocation of communities that are no longer viable in their current location. Equally, there will be a need for changes to the ways in which energy is produced and this may involve changing coal-fired power stations to gas-fired power or the construction of new power plants using renewables such as wind farms, for example. Typically, such developments take several years to progress from inception to completion. There is a view that action needs to be taken immediately and, therefore, the extensive planning consultations previously enjoyed may need to be shortened in the drive to reduce carbon emissions and thereby mitigate the effects of global warming and climate change.

Discussion point

List some of the risks involved when conducting business in another country.

10.7 Developing an international strategy

A decision to develop a global profile must be accompanied by a well planned and executed strategy. According to Griffin and Putstay (2007) there are five independent steps that should be undertaken. See also Figure 10.2.

1. Develop a mission statement. This clarifies the property developer's purpose, values and directions and is a means of communicating with internal and external constituents and stakeholders about the company's strategic direction.

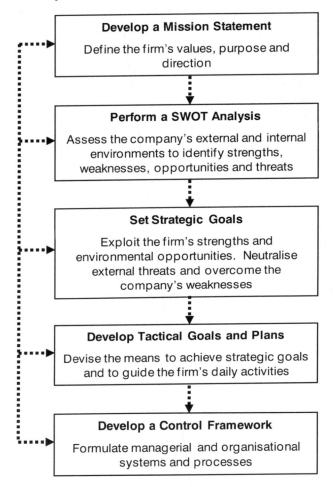

Figure 10.2 Steps in international strategy formation (Source: Griffin and Putstay, 2007)

2. Undertake environmental scanning and a SWOT (strengths, weaknesses, opportunities and threats) analysis. An environmental scan is a systematic collection of data about all elements of the property developer's external and internal environments, including markets, regulatory issues, competitor's actions, production costs and labour productivity.

3. Set strategic goals. These are the major objectives the developer wishes to accomplish through pursuing a particular course of action. Importantly, they should be measurable, feasible and time restricted.

4. Develop specific tactical goals or plans that focus on the details of implementing the property developer's strategic goals.
5. A control framework is required. This is the set of organisational and managerial processes that keep the property developer moving towards its strategic goals.

10.8 Examples of international property development

There are many examples of international property developments being undertaken. These examples feature different types of development operating in different regional locations. The first example is a hotel and resort development proposed by the Trump Organisation in the Middle East; this region has experienced rapid property development over the last few decades as result of the massive oil revenues. The region is now seeking to promote itself as a tourist destination to attract people to the area. The second example is a redevelopment of an old textile mill in the marketplace area of Lodz in Poland.

The Palm Jumeirah, Trump International Hotel and Tower, Nakheel, United Arab Emirates

The Trump Organisation is an excellent example of an international property developer. Trump is a US-owned company operating with this development in the Middle East. They recently revealed details about the design and their increased involvement in the United Arab Emirates (UAE). The Trump Organisation is planning to develop mixed use developments and condo hotels in the Middle East. The Trump International Hotel and Tower, The Palm Jumeirah, is the initial development in Nakheel and is a joint venture in the Middle East that includes exclusive rights for nineteen countries in the Middle East region and seventeen major brands. It is also the first UAE property in the portfolio of Nakheel Hotels and Resorts, Nakheel's hotel and resort investment company, which was launched in February 2006. The organisation's architects and designers engage closely with Nakheel Hotels and Resorts on the design, to utilise local knowledge and expertise, and the results of the partnership have been released.

The US$ 600 million property, The Palm Jumeirah, is a forty-eight-storey mixed-use hotel and residential building, anchoring the trunk of the five-by-five kilometre, manmade palm-tree shaped island that lies off the coast of Dubai. It is the first of three such islands to be built in Dubai. The Palm

Jumeirah will be a premier resort, a mixed-use development to spread risk, and will offer beachfront hotels, residences, retail and leisure. The design features a split, linked tower – an open-core design that minimises shadows – constructed with stainless steel, glass and stone. The property is intended to be a striking landmark to attract investors and buyers. The property's tall, slender design allows for a linear view through the building to the top of the island and provides panoramas of the island, Dubai and the Arabian Gulf, with all rooms benefiting from a sea view, again to attract international buyers and investors. Although the development will meet a number of international standards in respect of quality and fit-out, the sustainability credentials of the development are not clear at present. The development is already promoted on the web yet no construction has commenced.

Manufaktura, Lodz, Poland

Lodz is the second city of Poland with a population of over 900,000 and a student population of 100,000. After Warsaw, Lodz is one of the most popular tourist destinations in Poland along with Krakow and Poznan. The 'golden age' of Lodz was during the industrial revolution and it has continued since then to embrace four cultures: Jewish, German, Russian and Polish. Its fame was expanded by Israel Poznanski who was the founder of the textile factory in 1852, where, by 1877, was employing more than 12,000 workers. Following the death of the founder, it fell into disrepair and the company eventually became bankrupt. The factory itself remained derelict awaiting a new use.

Manufaktura is a 27-hectare district located in the heart of the city, where the objective was to bring the area back to life by renovating and refurbishing 90,000 m² of nineteenth-century historical building. Overall, 45,000 m² of brick facade was renovated to create a link between the past and the future, and the project incorporated social sustainability and the reuse of embodied energy in the existing construction materials. Most importantly, the project was designed to respect the local community and involve it throughout the redevelopment process. The project aimed to create a dialogue between the four cultures and also bring other social cultures into the development.

Table 10.1 shows the four major components used in the project. The main project will include:

- Twelve-screen cinema complex and a 3D IMAX (Cinema City)
- 1,000 m² children's playground (Kinderplaneta)
- Mixed education and entertainment building (Eksperymentarium)

Plate 10.1   Property development in Manufaktura, Poland. Reproduced with permission of King Sturge

- Gymnasium and fitness centre
- Disco and casino
- Bowling alley with sixteen lanes
- Open-air cinema
- Internal tram system
- Restaurants and gardens

The property development is a joint venture company pioneered by APSYS, jointly funded and financed by Paris Orleans, Fonciere Euris and

Table 10.1   The four major components of the Manufaktura project

Culture	8,000 m²
Business (hotel/office)	20,000 m²
Retail	101,000 m²
Leisure	21,000 m²
Total gross lettable area	150,000 m²

Euro-Hypo Bank, and is an example of financial backing from one country funding property development and commercial enterprise in the retail and recreational sector in another country.

(Note that the information for this case study was obtained from *European Property Sustainability Matters* published by King Sturge, London, and is reproduced with their permission.)

Reflective summary

The extent and scope of world trade has increased substantially in the twentieth century and economic events in one country do impact economically on other nations. Furthermore, a considerable amount of investment occurs in property across national borders. Globalisation has homogenised market expectations to a substantial degree in many areas, including property. For example, the standards and specifications in respect of the commercial office, international hotel and retail property sectors are highly aligned in the developed countries.

Property developers have embraced the concept of global business with many companies adopting an international perspective. This has presented many opportunities for progressive property developers who have expanded their horizons. Globalisation has provided opportunities for property developers to meet increased market requirements and demands in all property sectors in the UK as well as providing new opportunities for UK developers to operate outside of the UK. Expansion of the EU and relaxation of trade barriers has helped promote increased activity in this respect. In addition, the growth of information technology and the internet has made the operation and management of the property development process much easier to undertake regardless of whether the developer is located in the UK or elsewhere. In some niche areas opportunities have arisen for commercial property development supporting the expansion of UK firms outside the UK and also in the residential property development sector to service the growing numbers of UK buyers seeking to purchase homes outside the UK but who prefer to deal with either English speakers or English companies. However, it is clear that buyers are becoming more used to the concept of buying and owning property outside their country of residence and as a result these buyers are becoming more confident of buying, not only 'off the plan' without seeing the property but also of buying in countries further afield.

Sound market research is a prerequisite for successful international property development and access to local knowledge and expertise is essential. Different countries offer different opportunities within the various market sectors for property development. For example, India, China, Russia, Brazil and Mexico are all forecast to experience strong growth within the coming decade.

Along with the opportunities there are risks and barriers to success. The key barriers are those related to language and cultural issues, local knowledge and practices, along with the benchmarking of appropriate standards of property development. Property developers must appreciate that subtle yet significant difference in the way business is done can make working outside of national boundaries a challenging experience at times. Property developers who ignore or underestimate cultural issues do so at their peril. Finally, a lack of local market knowledge and market volatility can lead developers to experience much lower returns and expose them to much higher risk than anticipated. Many regions promote a 'buy local' culture so it is important to consider the culture and environment of the area surrounding the property development. Employing local workers where possible will partly address these concerns, and for large projects a public relations expert may be employed to ensure interested stakeholders are kept up to date with important facets of the development.

Case study

The case study provided here is a purpose-built office building for an owner–occupier in Melbourne, Australia, completed in 2006.

Council House 2 or CH2, is a ten-storey office building accommodating around 540 people, with ground floor retail spaces and underground parking. It was opened in August 2006.

The vital statistics are as follows:

- Gross floor area (GFA): 12,536 m² comprising:

    - 1,995 m² GFA basement areas
    - 500 m² net lettable area (NLA) – ground floor retail

Plate 10.2   CH2 external elevation, Swanston Street, Melbourne

- 9,373 m² total NLA
- 1,064 m² GFA – typical floor

- Bike spaces: 80
- Showers for cyclists: 9
- Car spaces: 20 plus one disabled space. The car park can be converted to office space or other uses

Total building cost

The total cost is Australian $51,045 million, which includes CH2 building costs, roadwork, upgrades to other buildings, professional fees, relocation costs, fit-out, art costs, footpaths, landscaping and other costs. CH2 building costs include:

- $29.9 million for the base building (2,334 $/m² or 58.5 per cent of cost).
- $11.3 million for sustainability features including a portion of the building cost of purge windows, light harvesting devices, precast ceilings, timber shutters, precast exhaust ducts, solar hot water collectors, photovoltaic cells, chilled water cooling system, shading screens, co-generation plant, air conditioning, and beams and slabs (884 $/m² or 22.1 per cent of cost).

- $2.8 million on education and demonstration including a portion of the cost of shower towers, multi-use water treatment plant, PCM (phase change material) modules, roof landscaping and chilled ceiling panels/beams (218 $/m² or 5.5 per cent of cost).
- $7.1 million on requirements specific to council use including a portion of the cost of vertical landscape, balconies, access floors, lift finishes, communication cabling, standby generator, security system and building automation system (553 $/m² or 13.9 per cent of cost).

Payback period

It is estimated that in 10 years time the sustainability features will have paid for themselves. Further benefits that could reduce this figure include:

- Healthier staff – less time lost to colds, flu and other illnesses.
- Increased workplace effectiveness.
- Less costs for public domain and infrastructure.
- The value of building as a guiding light in sustainable building.

How it works

CH2 is designed to reflect the planet's ecology, which is an immensely complex system of interrelated components. Just as it is impossible to assess the role of any part of this ecology without reference to the whole, CH2 comprises many parts that work together to heat, cool, power and water the building, creating a harmonious environment. For example, in nature, dark colours absorb heat and hot air rises. Accordingly, CH2's north facade comprises ten dark-coloured air extraction ducts that absorb heat from the sun, helping stale air inside rise up and out of the building. The south facade comprises light-coloured ducts that draw in fresh air from the roof and distribute it down through the building. Occupants can control the flow of this 100 per cent fresh air to their work spaces by floor vents. Louvres made from recycled timber shade the west facade. Energy from photovoltaic roof panels powers the louvres, which move according to the position of the sun. These features combine to create a controlled and healthy climate.

About 100,000 litres of black (toilet) water a day is extracted from the street sewer. A city sewer usually holds 95 per cent water, a burden on the system and a waste of water. The sewage, along with any generated on site, is put through a multi-water treatment plant that filters out the water and sends solids back to the sewer. The extracted water is treated through a micro-filtration system to create water suitable for all non-drinking uses.

Some of the recovered water supplies CH2's water cooling, plant watering and toilet flushing needs. The rest is used in other council buildings, city fountains and plants. More water is saved through recycling water from the fire-safety sprinkler system and

Plate 10.3   CH2 as the world's greenest office building

from rainwater. CH2 has received a rating of six stars from the Green Building Council of Australia. The building is a world leader in green design and was assessed via the Green Star method earning six Green Stars, which represents world leadership in office building design. For further information on this building see http://www.melbourne.vic. gov.au.

Appendix 1: Example mission statement on CSR

Sustainability – our vision

Lend Lease is a leading retail and residential property group integrated with strong investment management and construction management businesses. Our vision is to become a sustainable organisation. Sustainability defines the way we do business now and into the future: it's a commitment that ranges from living our corporate values and being incident and injury free through to the consistent delivery of environmental solutions and long-term prosperity for all stakeholders.

We recognise the legacy of our activities and we are accountable for what we take, what we create and what we leave behind. We have defined what sustainability means for each of our businesses and this helps guide our employees and partners towards realising the vision:

	How we operate	What we produce and influence
Social sustainability	• We have eliminated work-related death and injury;	• We engage with stakeholders to understand how our work affects them so we can enhance the projects we deliver, own and operate;
	• We engage with our stakeholders to understand how our business affects them;	• We create lasting linkages between the places we create and the people who use or are influenced by them; and

continued...

	How we operate	*What we produce and influence*
	• We are committed to working ethically, and with professionalism and integrity;	• We continuously contribute to educating and influencing the communities we work in to live sustainably.
	• We have a healthy culture aligned with our values;	
	• We have a diverse and flexible workplace;	
	• We actively contribute to the communities in which we operate; and	
	• We continuously inform and encourage our people to live and work sustainably	
Economic sustainability	• We demonstrate world leading sustainability practice while delivering the economic performance expected of us;	• We use property as a catalyst for regional economic development;
	• We positively contribute to the economic development of the regions in which we operate; and	• We create and facilitate skilling and employment for local people;
	• We continue to operate as an economically healthy business.	• We support and promote local and small business; and
		• We continue to produce economically sustainable products.
Environmental sustainability	• Our products are designed to exceed the best environmental standards and lead the market in environmental design and/or performance; and	We are energy, water and waste neutral in • What we design; • What we construct; • What we develop;
	• The buildings we occupy are neutral for energy, water and waste and best practice for indoor environment and air quality.	• For the communities we build and influence; • For what we manage; and • For what we own.

Bibliography

Chapter 1: introduction

Barras, R. (1994) Property and the economic cycle: building cycles revisited, *Journal of Property Research*, 11, 183–97.

Dixon, T., Thompson, B., McAllister, P., Marston, A. and Snow, J. (2005) *Real Estate and the New Economy*, Blackwell Publishing, London.

English Partnerships http://www.englishpartnerships.co.uk/

KPMG (2007) *Real Estate Investment Trusts (REITs)*, London.

Reed, R.G. (2007) *Valuation of Real Estate*, Australian Property Institute, Canberra.

Royal Institution of Chartered Surveyors (1994) *The Mallinson Report. Report of the President's Working Party on Commercial Property Valuations*, Royal Institution of Chartered Surveyors, London.

Royal Institution of Chartered Surveyors (RICS) http://www.rics.org

Scottish Enterprise http://www.scottish-enterprise.com/sedotcom_home/about_se.htm

University of Aberdeen and Investment Property Databank (1994) *Understanding the Property Cycle*, Royal Institution of Chartered Surveyors, London.

Welsh Development Agency http://www.wda.co.uk/index.cfm/wda_home/index/en2

Chapter 2: Land for development

Ball, S. and Bell, S. (1991) *Environmental Law*, Blackstone Press, London. Chesterton Research (July 1993) City Office Markets: The National

British Urban Regeneration Association http://www.bura.org.uk/

CABE (Commission for Architecture and the Built Environment http://www.cabe.org.uk/

Communities and Local Government http://www.communities.gov.uk

Communities and Local Government http://www.communities.gov.uk/pub/769/commercialandIndustrialFloorspaceandRateableValueStatistics2005PDF 6100Kb_id1163769.pdf; UK government statistics on commercial and industrial floor space and rateable value.

Department for Transport http://www.dft.gov.uk

Department for Transport – Highways Agency http://www.highways.gov.uk

English Partnerships http://www.englishpartnerships.co.uk/

Havard, T. (2002) *Contemporary Property Development*, RIBA Enterprises, London.

Keeping, M. and Shiers, D. (2004) *Sustainable Property Development*, Blackwells Science Limited, Oxford.

Renewal.net http://www.renewal.net/

Rogers, R.G. (1999) *Towards an Urban Renaissance. Final Report of the Urban Task Force*, Taylor & Francis, London.

Rogers, R. and Power, A. (2000) *Cities for a Small Country*, Faber, London. .

Syms, P. (2002) *Land, Development and Design*, Blackwells Science Limited, Oxford.

Chapter 3: development appraisal and risk

Baum, D. and Mackmin, D. (1989) *The Income Approach to Property Valuation*, 3rd edn, Routledge & Kegan Paul, London.

Byrne, P. (in press) *Risk and Uncertainty and Decision-Making in Property Development*, 2nd edn, E & FN Spon, London.

International Valuation Standards Committee (2005) *International Valuation Standards*, 7th edn, London.

Royal Institution of Chartered Surveyors (1994) *The Mallinson Report. Report of the President's Working Party on Commercial Property Valuations*, Royal Institution of Chartered Surveyors, London.

Chapter 4: development finance

CB Richard Ellis (2007) *CB Hillier Parker Rent and Yield Monitor*, London.

Darlow, C. (1989) Property development and funding, in *Land and Property Development: New Directions* (ed. R.Grover), E & FN Spon, London, pp. 69–80.

Dixon, T., Thompson, B., McAllister, P., Marston, A. and Snow, J. (2005) *Real Estate and the New Economy*, Blackwell Publishing, London.

English Partnerships http://www.englishpartnerships.co.uk

IPD (2007) *IPD UK Annual Property Index 2006*, Investment Property Databank Ltd, London.

Kolbe, P.T., Greer, G.E. and Rudner, H.G. (2003) *Real Estate Finance*, Dearborn Real Estate Education, Chicago.

KPMG (2007) *Real Estate Investment Trusts (REITs)*, London.

Office for National Statistics (2007a) *Focus on Consumer Price Indices*, London.

Office for National Statistics (2007b) *Investment by Insurance Companies, Pension Funds and Trusts*, London.

Reed, R.G. (2007) *Valuation of Real Estate*, Australian Property Institute, Canberra.

Rodney, W. and Rydin, Y. (1989) Trends towards unitisation and securitisation, in *Land and Property Development: New Directions* (ed. R. Grover), E & FN Spon, London, pp. 81–94.

Royal Institution of Chartered Surveyors (2007a) *REITs on the rise – Has a Revolution Begun?* Royal Institution of Chartered Surveyors, 17 January, London.
Royal Institution of Chartered Surveyors (2007b) *Teething Trouble for New UK REITs*, Royal Institution of Chartered Surveyors, 16 May, London.

Chapter 5: Planning

Ball, S. and Bell, S. (1991) *Environmental Law*, Blackstone Press, London.
Barker Review of Land Use Planning Final Report – Recommendations (2006) December 2006. www.barkerreviewofplanning.org.uk.
Communities and Local Government (2007) *Planning for a Sustainable Future: White Paper.* .
Cullingworth, B. and Nadin, V. (2006) *Town and Country Planning in the UK*, 14th edn, Routledge, Abingdon.
Moore, V. (1993) *A Practical Approach to Planning Law*, 3rd edn, Blackstone Press, London.
Newman, P. and Thornley, A. (1996) *Urban Planning in Europe. International Competition, National Systems and Planning Projects*, Routledge, London.
Planning Advisory Service (PAS) http://www.pas.gov.uk
Planning Aid http://www.planningaid.rtpi.org.uk
Planning Inspectorate http://www.planning-inspectorate.gov.uk/pins/index.htm
Planning Officer's Society http://www.planningofficers.org.uk
Planning Portal http://www.planningportal.gov.uk
Planning Resource http://www.planning.haynet.com
Royal Institution of Chartered Surveyors http://www.rics.org
Royal Town Planning Institute http://www.rtpi.org.uk
Tewdwr-Jones, M. and Williams, R.H. (2001) *The European Dimension of British Planning*, Spon, London.
The Planning Inspectorate (2007) Table 1.2 Planning Appeals, http://www.planning-inspectorate.gov.uk
Town and Country Planning Act 1990 http://www.opsi.gov.uk/ACTS/acts1990/Ukpga_19900008_en_1.htm
Town and Country Planning Act (Scotland) 1997 http://www.opsi.gov.uk/acts/acts1997/1997008.htm
Town and Country Planning Association http://www.tcpa.org.uk
United Nations (1987) *Our Common Future: Report of the World Commission on Environment & Development.* General Assembly Resolution 42/187, 11 December 1987.

Chapter 6: Construction

Construction Industry Council (1994) *The Procurement of Professional Services – Guidelines for the Value Assessment of Competitive Tenders.* http://www.cic.org.uk
Construction Industry Council (2002) *A Guide to Project Team Partnering*, 2nd edn, April 2002. http://www.cic.org.uk

Construction Industry Council (2005) *Selecting the Team.* http://www.cic.org.uk

Egan, J. (1998) *Rethinking Construction* (The Egan Report), Department of the Environment, Regions and Transport, HMSO, London.

Kelly, J., Morledge, R. and Wilkinson, S. (eds) (2002) *Best Value in Construction*, Blackwell Publishing, Oxford.

Morledge, R., Smith, A. and Kashiwagi, D. T. (2006) *Building Procurement,* Blackwell Publishing, Oxford.

Latham, M. (1994) *Constructing the Team* (The Latham Report), Department of the Environment, HMSO, London.

National Audit Office (2001) *Modernising Construction*, Report by the Comptroller and Auditor General, HC 87 Session 2000–2001, 11 January 2001. The Stationery Office, London.

ODPM (2002) *Green Private Public Partnerships*, Department for Transport/ Department for Environment, Food and Rural Affairs/Office of Goverment Commerce, Norwich.

Wood, G.D. and Ellis, R.C.T. (2005) Main contractor experiences of partnering relationships on UK construction projects, *Construction Management and Economics*, 23(3), March, pp. 317–25.

Chapter 7: Market research

Appraisal Institute (2001) *The Appraisal of Real Estate*, The Appraisal Institute, Chicago.

Australian Property Institute (2007) *The Valuation of Real Estate* (ed. R. Reed), The Australian Property Institute, Canberra.

Fanning, S. (2005) *Market Analysis for Real Estate*, The Appraisal Institute, Chicago.

Fanning, S., Grissom, T. and Pearson, T. (1994) *Market Analysis for Valuation Appraisals*, The Appraisal Institute, Chicago.

Chapter 8: promotion and selling

Adams, D., Watkins, C. and White, M. (2005) *Planning, Public Policy and Property Markets*, Blackwell Publishing, Oxford.

Dixon, T., Thompson, B., McAllister, P., Marston, A. and Snow, J. (2005) *Real Estate and the New Economy*, Blackwell Publishing, London.

Forlee, R. (2006) *An Intelligent Guide to Australian Property Development*, John Wiley & Sons, Brisbane.

Klein, S.D., Reilly, J.W. and Barnett, M. (2004) *Real Estate Technology Guide*, Dearborn Real Estate Education, Chicago.

Reed, R.G. (2007) *Valuation of Real Estate*, Australian Property Institute, Canberra.

Chapter 9: Sustainable development

Better Buildings – Better Lives. Summary of the UK Government on Sustainable BRE. http://www.bre.co.uk/envprofiles 2007 (Accessed 24 May 2007).

Brinkman, R.L. (1999) The dynamics of corporate culture: conception and theory, *International Journal of Social Economics*, 26(5), 674–94.

Buildings Task Group Report (2006) http://www.berr.gov.uk/files/file15151.pdf

Burke, L. and Logsdon, J.M. (1996) How corporate social responsibility pays off, *Long Range Planning*, 29(4), pp. 495–502.

DEFRA (Department of Environment Food and Rural Affairs) http://www.defra.gov.uk/environment/sustainable/index.htm

Clark, C.E. (2000) Differences between public relations and corporate social responsibility: An Analysis, *Public Relations Review*, 26(3), pp. 363–80.

Environment Agency (2006) *Sustainable Construction Position Statement 2006*, http://www.environment-agency.gov.uk/aboutus/512398/289428/654938/?version=1&lang=_e&lang=_e (Accessed 24 May 2007).

Esrock, S.L. and Leichty, G.B. (1998) Social responsibility and corporate web pages: self presentation or agenda setting? *Public Relations Review*, 23(3), pp. 305–19.

European Union (2002) *Communication from the Commission Concerning Corporate Social Responsibility: A Business Contribution to Sustainable Development*, 2 July. http://trade.ec.europa.eu/doclib/docs/2006/february/tradoc_127374.pdf

Frankental, P. (2001) Corporate social responsibility – a PR invention? *Corporate Communications*, 6(1), pp. 18–23.

Friedman, A.L. and Miles, S. (2001) Socially responsible investment and corporate social and environmental reporting in the UK: an exploratory study, *British Accounting Review*, 33, pp. 524–48.

International Sustainable Development http://www.sustainable-development.gov.uk//international/index.htm

IPCC (2007) *Inter-Governmental Panel on Climate Change Working Group II Fourth Assessment Report*, 4 April, http://www.ipcc.ch/SPM6avr07.pdf

King Sturge (2007) *European Property Sustainability Matters*, King Sturge, London.

Lantos, G.P. (2001) The ethicality of altruistic corporate social responsibility, *Journal of Consumer Marketing*, 19(3), pp. 205–30.

Lewis, L. and Unerman, J. (1999) Ethical relativism: a reason for differences in social reporting? *Critical Perspectives on Accounting*, 10, pp. 521–47.

Maignan, I. and Ferrell, O.C. (2003) Nature of corporate responsibilities: perspectives from American, French and German consumers, *Journal of Business Research*, 56, pp. 55–67.

Milne, M.J. and Chan, C.C. (1999) Narrative corporate social disclosures: how much of a difference do they make to investment decision-making? *British Accounting Review*, 31, pp. 439–57.

O'Riordan, T. (2000) *Environmental Science for Environmental Managers*, Prentice Hall, Harlow.

PWC (2003) *5th Annual Global CEO Survey*, PricewaterhouseCoopers.

Rao, S., Yates, A., Brownhill, D. and Howard, N. (2003) *ECOHOMES: The Environmental Rating for Homes*. BRE, Watford.

RICS (2007) *A Green Profession?* RICS, London.

Sustainable Development Commission http://www.sd-commission.org.uk/
Sustainable Development UK http://www.sustainable-development.gov.uk/
UK Government Sustainable Development Action Plan (2006) http://www.
    communities.gov.uk/pub/444/ODPMsSustainableDevelopmentActionPlan_
    id1164444.pdf
UNFCCC (2007) *United Nations Framework Convention on Climate Change*, http://
    unfccc.int/resource/docs/convkp/kpeng.html
WCED (World Commission on Environment and Development) (1987) *Our Common
    Future*, Oxford University Press, Oxford.
Williams, S.M. and Pei, C.A. (1999) Corporate social disclosures by listed companies
    on their web sites: an international comparison, *The International Journal of
    Accounting*, 34(3), pp. 389–419.
World Business Council for Sustainable Development (2000) *Making Good Business
    Sense*, http://www.wbcsd.org

Chapter 10: International practice

Bröchner, J., Rosander, S. and Waara, F. (2004) Cross-border post-acquisition
    knowledge transfer among construction consultants, *Construction Management
    and Economics*, 22, pp. 421–7.
Bromiley, P. and Cummings, L. (1993) Transition costs in organisations with trust,
    Research and Negotiation in Organisations, 5, pp. 219–47.
Dehesh, A. and Pugh, C. (2000) Property cycles in a global economy, *Urban Studies*,
    37(13), pp. 2581–602.
Encyclopaedia Britannica http://www.britannica.com
ENR (Engineering News-Record) (2003) The Top International Contractors,
    Engineering News-Record, 251(8), pp. 36–41.
Ghoshal, S. and Nohria, N. (1993) Horses for courses: organizational forms for
    multinational corporations, *Sloan Management Review*, 34(2) (Winter),
    pp. 23–35.
Guy, S. and Henneberry, J. (2000) Understanding urban development processes:
    integrating the economic and the social in property research, *Urban Studies*, 37,
    pp. 2399–416.
Griffin, R.W. and Putstay, M.W. (2007) *International Business*, Pearson Prentice Hall,
    Sydney.
Hailia, A. (2000) Why is Shanghai building a giant speculative property bubble?
    International Journal of Urban and Regional Research, 37(12), pp. 2241–56.
Hill, R.C. and Kim, J.W. (2000) Global cities and development states: New York,
    Tokyo and Seoul, *Urban Studies*, 37(12), pp. 2167–95.
Jin, X. and Ling, Y.Y.L. (2005) Constructing a framework for building relationships
    and trust in project organizations: two case studies of building projects in China,
    Construction Management and Economics, 23, pp. 685–96.
King Sturge (2007) *European Property Sustainability Matters*, King Sturge, London.
Mansfield, J.R. and Royston, P.J. (2007) Aspects of valuation practice in Central and
    Eastern European economies, *Property Management*, 25(2) , pp 150–63.

McAllister, P. (2000) Is direct investment in international property markets justifiable? *Property Management*, 18(1), pp. 25–33.

Nappi-Choulet, I. (2006) The role and behaviour of commercial property investors and developers in French urban regeneration: the experience of the Paris region, *Urban Studies*, 43(9), pp. 1511–35.

Pheng, L.S., Hongbin, J. and Leong, C.H.Y. (2004) A comparative study of top British and Chinese international contractors in the global market, *Construction Management and Economics*, 22, pp. 717–31.

Raftery, J., Pasadilla, B., Chiang, Y.H., Hui, E.C.M. and Tang, B.S. (1998) Globalization and construction industry development: implications of recent developments in the construction sector in Asia, *Construction Management and Economics*, 16, pp. 729–37.

Sha, K. and Lin, S. (2001) Reforming China's construction state-owned enterprises, *Building Research and Information*, 29(4), pp. 270–6.

Sirmans, C.F. and Worzala, E. (2003) International direct real estate investment: a review of the literature, *Urban Studies*, 40(5–6), pp. 1081–114.

World Trade Organization http://www.wto.org

Wolfe, D.A. and Gertler, M.S. (2004) Clusters from the inside and out: local dynamics and global linkages, *Urban Studies*, 41(5–6), pp. 1071–93.

Index